Interdisciplinary Statistics

STATISTICS for FISSION TRACK ANALYSIS

T0174160

CHAPMAN & HALL/CRC
Interdisciplinary Statistics Series

Series editors: N. Keiding, B. Morgan, T. Speed, P. van der Heijden

Interdisciplinary Statistics

STATISTICS for FISSION TRACK ANALYSIS

Rex F. Galbraith

University College London, UK

CRC Press
Taylor & Francis Group
Boca Raton London New York

CRC Press is an imprint of the
Taylor & Francis Group, an **informa** business
A CHAPMAN & HALL BOOK

CRC Press
Taylor & Francis Group
6000 Broken Sound Parkway NW, Suite 300
Boca Raton, FL 33487-2742

First issued in paperback 2019

© 2005 by Taylor & Francis Group, LLC
CRC Press is an imprint of Taylor & Francis Group, an Informa business

No claim to original U.S. Government works

ISBN-13: 978-1-58488-533-7 (hbk)
ISBN-13: 978-0-367-39279-6 (pbk)

Library of Congress Cataloging-in-Publication Data

Catalog record is available from the Library of Congress

**Visit the Taylor & Francis Web site at
http://www.taylorandfrancis.com**

**and the CRC Press Web site at
http://www.crcpress.com**

Contents

List of tables

List of figures

Preface

In the 1960s fission track dating emerged as an alternative dating tool for geological time scales. It complemented other methods based on radioactive decay, such as potassium-argon dating, and it applied to older time scales than did radiocarbon dating — millions of years, rather than hundreds or thousands. In the 1970s it was found that fission track ages were sometimes younger than expected because of the effect of natural heating of a sample while it was buried. In the 1980s these effects began to be quantified and gave fission track analysis a new purpose: the reconstruction of geological thermal histories. It is now a major technique used by the oil exploration industry and in a variety of geological studies.

Fission track analysis has several special features. Firstly, unlike other thermal history tools, it can provide information on both times and temperatures. Secondly, it is inherently statistical in nature. Fission tracks are created over geological time and subjected to heating and other processes that, to a large extent, characterise the statistical distributions of their numbers, lengths and orientations. Thirdly, there is a simple stochastic model, the line segment model, that provides a rigorous framework for quantitative fission track analysis and a sound basis for scientific modelling.

This book is primarily about the line segment model and its practical implications for statistical analysis and modelling. It is based largely on my collaborations with Geoff Laslett over the last 20 years and with other scientists in this field. Much of the material is published in diverse sources and the book aims to bring it together with a coherent notation and terminology. I have tried to relate the underlying mathematics to empirical data and to provide explicit formulae where possible. I have also tried to emphasise practical interpretation in the context of real examples. Chapter 1 gives a general introduction to the subject and to the physical basis of the line segment model. Chapter 2 introduces the line segment model and derives probability distributions of numbers of tracks that intersect a plane. Chapters 3 and 4 are concerned with the theory and practice of fission track dating using the external detector and population methods and with the statistical analysis of the resulting data. Chapters 5 and 6 are about statistical modelling of mixed fission track ages, giving both theory and examples.

The line segment model is also used to analyse a number of geometric and observational effects inherent in measurements of fission track lengths and orientations. Chapter 7 derives theoretical probability distributions relevant to various length and angle measurements, while Chapter 8 deals with empirical features of such measurements. Detailed fitting of annealing models and thermal histories are beyond the scope of the book, but their statistical basis is discussed briefly in Chapter 9.

This book is aimed at various groups of research workers and students: those in fission track laboratories in universities and in oil companies, researchers and students in geology, geochronology and related fields — and also in applied probability and statistics, especially those working in the physical and geo-sciences, as an introduction to this field of application. Not all of these groups will be equally familiar with the mathematics or with the statistical concepts used. I have tried to make the text as widely accessible as possible by confining the main mathematical derivations to Chapters 2 and 7 and to a few other separate sections, and by adding verbal explanations. In parts the mathematical level is up to third year undergraduate, but many sections are at a lower level, including Chapters 3 and 4. I have also tried to make each chapter as self-contained as possible. This has resulted in some repetition but, I hope, has improved comprehensibility. For those less familiar with probability and statistics, I have added some notes in an appendix.

I owe debts of gratitude to many people. First to my friend and colleague Geoff Laslett, who is both an originator and co-author of much of the material in this book, and who contributed to some early drafts. Second, to Paul Green, who introduced me to this subject in 1980 and who has inspired me to continue my involvement with it ever since. Others who contributed to my early experience include, in order of fission track age, Chuck Naeser, Ian Duddy, Andrew Gleadow and Tony Hurford. I have benefited also from all in the fission track laboratories at the University of Melbourne, Geotrack, University College London, and from many others who have been immodest enough to show me their data. I am especially grateful to the whole fission track community for being simultaneously sympathetic and argumentative.

There is now a considerable literature on both methodology and applications of fission track analysis. I have tried to cite my sources, but I have not attempted a balanced review of the literature. This is of course reprehensible and I apologise to the many who have contributed to this field, but whose work I have not cited. My excuses are (a) that I am not the right person to do this and (b) that this is really a book about statistics.

I thank several people for kindly commenting on early drafts of parts of this book, including Trevor Dumitru, Nancy Naeser, Rodney Coleman, Malcolm Clark and Tony Hurford. I thank Jane Galbraith and Paul Green for critical comments on the whole text, Ina Dau for assistance with LaTeX and postscript programs and Roddy Kennard, Geoff Laslett, Brigid Janvier and Caroline Roaf for hospitality in Australia and France.

Rex Galbraith
April 2005

CHAPTER 1

Introduction

1.1 What are fission tracks?

Fission tracks are trails of damage in the crystal structure of a mineral, caused by the fissioning of uranium atoms. Several minerals, including apatite, zircon and sphene, typically contain trace uranium impurity incorporated at crystallisation. Over a long enough time, a small proportion of the uranium atoms, specifically the isotope ^{238}U, undergo spontaneous fission, where the atom splits into two parts that move rapidly in opposite directions, creating a long thin region of damage. Newly formed tracks in apatite are about 17 microns long and 0.01 microns wide. Tracks so formed are known as *spontaneous* tracks, or more correctly, spontaneous *fission* tracks. They are sometimes also called *fossil* tracks because they were formed millions of years in the past.

Fission tracks can also be created artificially by irradiating a mineral specimen with thermal neutrons in a nuclear reactor. These cause a small proportion of the isotope ^{235}U to undergo fission and similarly create tracks, which are known as *induced* tracks. Theories of the formation of fission tracks and the early history of the subject are described in the definitive textbook by Fleischer, Price and Walker (1975).

1.2 How are they observed?

Fission tracks are normally revealed by chemical *etching*. A mineral grain is ground and polished to expose a flat surface inside the crystal. It is then immersed in a chemical etchant that preferentially attacks the regions of damage, widening them and making them visible under an optical microscope. Tracks in apatite etch readily in 20 or 30 seconds when immersed in dilute nitric acid. Tracks in zircon and other minerals take longer. The etchant also attacks the mineral itself so if it is etched for too long the shapes of the tracks become indistinct.

Figure 1.1 is a photograph of etched spontaneous fission tracks in an apatite crystal. A plane interior surface, parallel to a prismatic face, has been polished and etched and tracks intersecting this plane are seen as darker regions. These are sometimes called *surface tracks* because they intersect the polished surface and sometimes *semi-tracks* because part of each original track (on average half) has been polished away. We will use the latter term in this book as we are mainly concerned with their mathematical and statistical properties. They have varying orientations and the dimensions seen in Figure 1.1 are those of

Figure 1.1 *Etched fission tracks in apatite. Tracks intersecting a polished surface (i.e., semi-tracks) are revealed by chemical etching. The larger dark etched regions are fractures in the crystal. The faint lines are polishing scratches. Photograph courtesy of Paul Green.*

their projections onto the polished plane. The etchant has also revealed some fractures in the crystal — the larger dark regions in Figure 1.1.

Figure 1.2 shows two more photographs of fission tracks at higher magnification. In the left panel the semi-track openings onto the polished plane can be seen as light-coloured slots aligned in the direction of the crystal's "c-axis", which is shown in the figure. There are also two fully *confined* tracks, labelled *a* and *b*, that have not intersected the polished surface. These have been revealed because they intersect a fracture in the crystal, the etchant having passed down the fracture and into the track. Such an observation is known as a *tincle* (*track-in-cleavage*). Track *b* is about 15 or 16 microns long and is at a low angle to the c-axis. Track *a* is rather shorter (8 or 9 microns) and at a high angle to the c-axis. The presence of tracks as short as this indicates that the crystal has been hot at some time in the past. Track *a* is also fatter than *b*. This is a feature of chemical etching in apatite. The etchant attacks the crystal damage faster in the direction of the c-axis than in the perpendicular direction.

The right panel of Figure 1.2 shows examples of two more confined tracks, labelled *c* and *d*. These have been revealed because they intersect etched semi-tracks and thus have themselves been etched. Such a confined track is called a *tint* (*track-in-track*).

An experienced observer using an optical microscope can count the number of semi-tracks intersecting a chosen area of surface and measure their lengths and orientations. By adjusting the focus of the microscope it is also possible

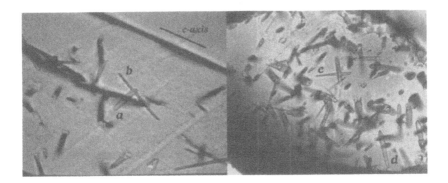

Figure 1.2 *Etched fission tracks in apatite. The semi-track openings on the polished plane are seen as light-coloured rectangular slots, and are aligned in the direction of the crystal's "c-axis". In the left panel there are two totally confined tracks (a and b) that have been etched because they intersect a fracture. These are tincles. In the right panel there are two tints (c and d), confined tracks that have been etched because they intersect semi-tracks. Photographs courtesy of Paul Green.*

to see etched tracks, or parts of tracks, a few microns below the surface. Similarly, the lengths and orientations of confined tracks may be measured. The discovery that totally confined tracks, fortuitously etched, could be seen and measured was a breakthrough in fission track analysis.

1.3 Why are they useful?

In a natural sample, the number of spontaneous tracks that intersect a surface depends on the concentration of uranium impurity and the time over which tracks have been forming. So if the former could be measured the latter might be deduced. This idea led to the development of *fission track dating* as a geochronological method — a dating method that is now well established in several geological contexts.

Furthermore, observations in both laboratory and geological conditions have shown that heating a mineral grain causes any fission tracks to shorten, or *anneal*, and with sufficient heat the tracks will disappear. There is a corresponding decrease in the density of tracks that intersect a polished mineral surface. The lengths of fission tracks and the density of their intersections on a surface are therefore important indicators of the thermal conditions to which the host mineral has been subject since formation. The mineral apatite is particularly useful in this context because the range of temperatures for which significant annealing takes place over geological time scales, 20–150°C, is lower than for other common minerals and corresponds to temperatures within the upper few kilometres of the earth's crust. It is therefore of special interest to model the behaviour of fission tracks in apatite under various thermal conditions.

Less is known about the annealing properties of fission tracks in zircon, but the corresponding range of temperatures is thought to be as high as 200–350°C, which makes zircon complementary to apatite and useful in other geological contexts.

1.4 Applications of fission track analysis

The reviews by Hurford (1991), Gallagher *et al.* (1998) and Dumitru (2000) identify a wide range of geological studies in which fission track analysis has been used. A major application is to the thermal evolution of sedimentary basins. The range of temperatures at which fission tracks in apatite shorten significantly over geological time overlaps that in which hydrocarbons are generated (Gleadow *et al.*, 1983, 1986). Apatite fission track analysis is now an important tool in oil exploration, where it can provide key information on both the cooling time and the maximum temperature a sample has experienced. A related application to ore mineral deposits is described by Arne (1992a).

A different use of fission track data is in provenance studies in sedimentary basins (e.g., Hurford and Carter, 1991) where sediments are derived from erosion of pre-existing rocks and contain apatite and zircon grains from the original source. Identification of different sources, obtained by dating these mineral grains, allows inferences about denudation rates and palaeodrainage directions, as well as providing a maximum depositional age for the sediment host.

A third application is to the structural evolution of mountain belts. Continental movements in the formation of mountains may result in crustal thickening, along with lateral and radial movements, surface uplift and erosion. There are different mechanisms that may produce surface uplift and it is hard to measure uplift rates, and changes in them, over geological time scales. Fission track analysis provides quantitative information on cooling and so has the potential to provide time and rate information on erosion, which in turn can put useful constraints on the uplift rates.

A variety of other applications arise in landscape evolution in non-mountain settings and in long-term continental denudation, where spatial and temporal information from fission tracks have different implications depending on the geological context. In Quaternary geochronology fission track analysis has been used to determine formation ages of volcanic flows and ashes, which then serve as marker dates for determining geological histories over time. Most such studies use ages obtained from zircon grains because they are relatively rich in uranium and more resistant to annealing, but some also date volcanic glasses and shards (e.g., Westgate, 1989; Bigazzi *et al.*, 1993).

A feature of most applications is that fission track data are used in conjunction with other geological tools and constraints. This greatly increases the usefulness of the data and the types of inferences that are possible. Indeed, the same fission track data and estimates may be used in quite different studies, and for different purposes.

1.5 Mathematical representation of fission tracks

Because fission tracks accumulate over time, the statistical distribution of their number, lengths and orientations reflects the whole thermal history experienced by the host crystal. This distribution may be understood and quantified in terms of a simple stochastic process to represent latent tracks inside a crystal — line segments of varying lengths and orientations located at points of a Poisson process in three-dimensional space. It turns out that this simple model is very realistic in the sense that it captures the essential qualitative and quantitative characteristics that are relevant to track measurements. Of course, tracks really have a thickness and a shape, but for the purposes of statistical modelling it is just their (idealised) lengths and orientations that matter.

In Chapter 2 we establish the line segment model and use it to derive a general formula for the expected number of tracks per unit area that intersect a plane. This derivation uses a simple counting argument plus a little three-dimensional geometry. An essential feature of this subject is that it is necessary to consider the three-dimensional process, even though observations may be made in one or two dimensions.

1.6 Fission track dating and provenance studies

A natural sample may have had a rather simple thermal history, where it has cooled relatively rapidly from a high temperature and subsequently remained cool until the present day. Then any fission tracks it contains will have formed since the time of cooling, and the "age" obtained by fission track dating will correspond to that time. Statistical methods for estimating this age, and validating it, are dealt with in Chapters 3 and 4. Standard fission track dating uses track counts but not length measurements. In the most efficient and most widely used experimental design, called the *external detector method*, an independent age estimate and its precision can be obtained for each single crystal in a sample.

Moreover, a sample may contain crystals of differing provenance, perhaps eroded from different locations and carried down a river to a common resting place, or perhaps crystals that have cooled at different times. Such a sample may contain a mixture of two or three (or more) different fission track ages. Mixtures of fission track ages may also arise from the effect of heat on crystal grains that have different chemical compositions. Unravelling information from such samples is the subject of Chapters 5 and 6. These chapters present some simple statistical models for mixtures of ages and show how to fit them using maximum likelihood estimation.

1.7 Thermal histories and track length distributions

To infer a thermal history — or to extract information about temperature as well as time — it is generally necessary not only to count tracks but also

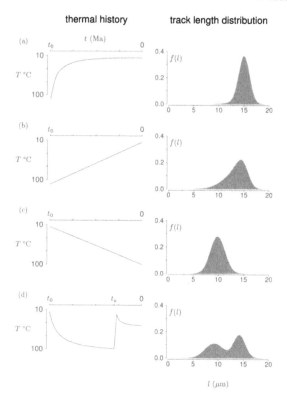

Figure 1.3 *Some thermal histories and corresponding track length distributions (from Laslett et al., 1994): (a) rapid cooling to the current temperature, (b) slow cooling to the current temperature, (c) slow heating to maximum temperature now, and (d) heating to a maximum temperature at time t_u followed by rapid cooling and subsequent heating. Here l denotes the full length in three dimensions and $f(l)$ is the relative number of tracks of length l.*

to measure their lengths. Figure 1.3 shows the frequency distribution of track lengths that might be produced by each of four hypothetical thermal histories. This figure uses the common conventions in geology of temperature increasing downwards on the y scale, as it does with depth below the earth's surface, and time before the present increasing from right to left on the x scale. These length distributions are based on models obtained from experimental observations and theoretical studies of fission track annealing kinetics (Laslett *et al.*, 1987; Duddy *et al.*, 1988).

In panel (a), representing rapid cooling to a low temperature, the resulting length distribution is unimodal and symmetrical with a high mean and small standard deviation. In panel (b), representing slow cooling, the length distribution has a lower mean, a higher standard deviation and is negatively skewed. Here the tracks that were formed earlier have experienced higher temperatures and are therefore shorter; these form the lower tail of the length distribution. Panel (c) represents slow heating to a maximum temperature at the present day and produces a fairly symmetrical length distribution but with a lower mean and higher standard deviation than in panel (a). For the thermal history in panel (d), tracks formed before time t_u have been heated to 100°C while those formed subsequently have only experienced a somewhat lower temperature, leading to the bimodal length distribution shown. Ideally it is of considerable interest to estimate the track length distribution and hence to infer the thermal history. However, this usually requires additional information because different thermal histories can produce quite similar length distributions.

1.8 Sampling by plane section

The process of polishing and etching a flat surface inside a crystal implies that, from a statistical point of view, tracks are selected by *plane section* sampling. The idea is illustrated in Figure 1.4, which gives a two-dimensional view, not to scale. A consequence of this sampling process is that the probability distributions of lengths and orientations of etched tracks, even if they could be measured exactly, differ from those of the underlying latent tracks.

In practice, several different types of measurements are made, including lengths and orientations of confined tracks, of semi-tracks and, of the projections of semi-tracks onto the plane of observation. Chapter 7 is about the theoretical properties of such measurements and derives formulae for their probability distributions in terms of the underlying line segment process. These distributions and their parameters are important both for understanding and for statistical modelling.

It can be seen from Figure 1.3 that the relation between thermal history and track length distribution is not straightforward. In addition, there are several observational factors that need to be understood in order to make correct inferences. These are addressed in Chapter 8. The next sections describe the physical basis of the line segment model.

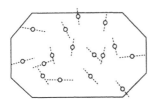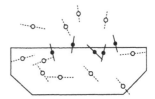

Figure 1.4 *Schematic illustration of sampling by plane section. Left panel: cross-sectional view of a crystal with latent tracks. Right panel: the same view after the crystal has been cut horizontally. The highlighted tracks are those that intersect the cut surface and will thus be etched.*

1.9 Initial formation of tracks

When a ^{238}U atom fissions spontaneously, two positively charged fission fragments are created. These mutually repel and smash their way through the lattice, knocking other atoms out of their lattice sites. The trajectories of the two fragments form a single fission track — a long thin region of damage in the crystal. The fission fragments vary in size, and their amounts of energy also vary, so that tracks may vary in length. There is no reason why any particular direction of the track may be preferred, nor is there any empirical evidence for this, so it is assumed that the orientations are random — that is, all directions are equally probable.

Spontaneous fission of a ^{238}U atom occurs with very low probability. For example, the proportion of atoms that fission spontaneously within 100 million years is less than 10^{-8}. Furthermore, spontaneous fission is intrinsic to an atom and is not dependent on the external environment. Consequently each atom's behaviour is independent of that of other atoms.

Similar considerations apply to fission induced by irradiating a crystal with thermal neutrons in a nuclear reactor. Occasionally a ^{235}U atom will absorb a thermal neutron, releasing a "binding" energy which is sufficient to cause fission. Thermal neutrons do not cause ^{238}U atoms to fission because in that case the binding energy released by absorbing a neutron is less than the critical energy for fission. Induced tracks formed by thermal neutron irradiation also vary in length and appear to have random orientations. The proportion p of ^{235}U atoms that are induced to fission depends on how many neutrons the crystal is irradiated with. This is determined by the neutron fluence set by the experimenter and in typical applications is such that p is less than 10^{-5}.

The processes of spontaneous and induced fission are physically unrelated, so that the numbers of each isotope within a crystal that undergo fission are statistically independent. Also, the ratio of the numbers of ^{235}U to ^{238}U isotopes is constant in nature, with a value of about 7.25×10^{-3}. This means that the concentration of the latter isotope in a sample may be deduced from a measurement of the former. Actually, the two isotopes have different half lives, so the isotopic ratio changes over time. The relevant ratio for our purpose is that at the time of measurement. However, because the two half lives are very long — 713 and 4510 million years, respectively — any change over a few thousand years is negligible. For example, in 1000 years the isotopic ratio will be multiplied by $1 - \epsilon$ where ϵ is less than 10^{-6}.

1.10 Shortening of tracks by heat

After track formation, the radiation damage remains in the crystal unless it is changed by environmental effects such as pressure, temperature or other natural processes. Over geological time, though, the main effect of practical relevance is that of heat. From field observations and laboratory experiments, it has been found that heating a mineral grain causes the radiation damage to repair — that is, the displaced atoms return to their original lattice sites.

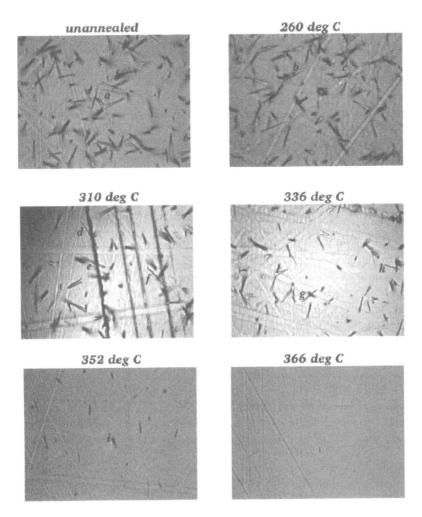

Figure 1.5 *Fission tracks intersecting a polished surface in six grains of Durango apatite. The c-axis is in the vertical direction in all panels. One grain was not heated and the other grains were heated for 1 hour at the temperatures shown. There are also some confined tracks identified by the letters a, b, c, d, e, f, g, h, i. Photographs courtesy of Paul Green.*

Any fission tracks will shorten, or *anneal*, and with sufficient heat the tracks will disappear.

Figure 1.5 shows photographs of induced fission tracks in six grains of apatite from Durango, Mexico. The grains were irradiated together, then one was kept aside (unannealed) and each of the others was heated for 1 hour at a different temperature. Most of these are semi-tracks intersecting the etched plane surface, but there are some confined tracks too, identified by letters.

The density of tracks intersecting the surface decreases with increasing temperature and those that have experienced higher temperatures are generally shorter. In the lower panels there are very few tracks remaining. Moreover, at the higher temperatures, the remaining tracks are mostly aligned in the direction of the c-axis, because the tracks at higher angles have annealed more quickly.

The confined tracks also tend to get shorter with more heat: a is the longest, b, c, d, e and f are somewhat shorter, g and h are rather short and track i is a very short tint perpendicular to the c-axis that can hardly be seen. These confined tracks tend to be at quite high angles to the c-axis. This is partly due to an observational phenomenon known as *orientation bias*, largely caused by the etching process. Because the etchant travels faster in the direction of the c-axis, a potential host fracture or semi-track will present a wider face in that direction, which is then more likely to be intersected by a track perpendicular to it. This same property also affects the shape of the etched tracks. Track b for example, is at a higher angle than track c and is consequently fatter.

Annealing of fission tracks in the mineral apatite has been studied extensively. It has been found that tracks in apatite repair from the ends inwards (Green *et al.*, 1986), so they do seem to retain their approximately linear shape under annealing, at least until shortly before the damage is repaired completely. Furthermore, in contrast to an unannealed sample, it has been observed that the track length distribution in annealed apatite is *anisotropic* (e.g., Green and Durrani, 1977; Green, 1981b; Donelick *et al.*, 1990 and Donelick, 1991). Tracks at a high angle to the crystallographic c-axis tend to be anneal (shorten) more quickly than those at a lower angle. Prior to annealing, the length distribution is nearly isotropic. As annealing progresses, not only does the mean length decrease, but also the anisotropy becomes more

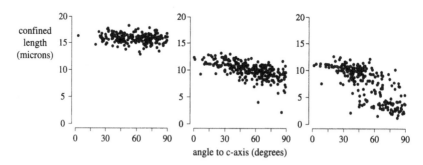

Figure 1.6 *Scatter plots of measured lengths and angles to the c-axis for three samples of induced horizontal confined tracks in Durango apatite. Left panel: unannealed; middle panel, heated for 1 hour at 300° C; right panel, heated for 1 hour at 343° C. The annealed tracks have an anisotropic length distribution and may be of two types: non-gapped tracks, whose lengths make up an anisotropic "envelope", and very short "gapped" tracks. These data are from the "Loaded Dog" experiments described in Chapter 8.*

marked. This is illustrated in Figure 1.6, using measured lengths and angles of horizontal confined tracks. The anisotropy appears to depend only on the angle of the track to the c-axis, and not on the angle of the projection of the track in the plane orthogonal to the c-axis.

There are other features seen in Figure 1.6. For example, the observed angles are not uniformly distributed over the range 0 to 90 degrees, and there are some very short measured lengths in the highly annealed samples. These are mainly due to observational factors — etching and geometrical biases — to be discussed in Chapter 8.

1.11 Properties of apatite

A major application of fission track analysis is to the estimation of geological thermal histories using apatite. This mineral has been well studied and has several useful properties.

Firstly, it is commonly occurring and it contains trace uranium impurity in appropriate concentrations. Secondly, the range of temperatures in which significant annealing takes place over geological time is lower than for other common minerals, such as sphene, zircon, mica and quartz. Furthermore, this temperature range (20–150°C) overlaps with those of geological processes of interest. Thirdly, apatite is optically transparent in thin section, so tracks are readily observed; and fourthly, unlike zircon and sphene, apatite does not retain alpha-recoil damage, which can change the crystal properties over time.

Elliot (1994) documents the structure and chemistry of apatites. They have the chemical formula $Ca_{10}(PO_4)_6X_2$, where X can be an F^- ion (fluorapatite), an OH^- ion (hydroxyapatite) or a Cl^- ion (chlorapatite). They occur as minor constituents of many igneous rocks, although some large igneous deposits do exist, e.g., on the Kola peninsula in the Russian Federation, and they are also present in most metamorphic rocks. Apatites also form the mineral component of bones and teeth, while less well crystallised forms (rock phosphates) occur in large deposits and provide most of the world's supply of phosphate for the fertiliser and chemical industries.

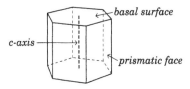

Figure 1.7 *The crystal structure of apatite, a first-order hexagonal prism. The two basal surfaces are regular hexagons and the six prismatic faces are rectangular (from Galbraith and Laslett, 1988).*

The basic crystal structure of apatite is a first-order hexagonal prism, with six prismatic faces and two basal surfaces (Figure 1.7). The central axis running between the two basal surfaces is called the crystallographic c-axis. The structure is very tolerant of ionic substitutions and, for example, Ca^{2+} ions can be partly or completely replaced by Ba^{2+}, Sr^{2+} or Pb^{2+} ions. Some lattice substitutions cause a lowering of the symmetry so that the unit cell may be doubled or slightly distorted from hexagonal. These changes in the symmetry often depend on the specific chemistry of the apatite and may affect its annealing properties.

1.12 Bibliographic notes

Fission track dating is said to have started with the dating of a natural mica by Price and Walker (1963), which was motivated by the first observation, through an electron microscope, of fission tracks in mica by Silk and Barnes (1959). More recently, Paul and Fitzgerald (1992) showed a transmission electron microscope image of an un-etched fission track in a fluorapatite. The scientific basis of fission track analysis was firmly established in the book by Fleischer, Price and Walker (1975). A shorter introduction to the physics of track formation and detection is given by Durrani and Bull (1987).

There are a number of very useful accounts of various aspects of the method and its applications. These include Naeser (1979a), Wagner (1979a, 1979b), Gleadow (1981), Naeser and Naeser (1984), Green et al. (1989a), Westgate (1989), Dumitru et al. (1991), Hurford (1991), Gallagher et al. (1998) and Dumitru (2000). The last three of these articles also review a range of applications and contain fairly extensive bibliographies. Papers reflecting a range of applications include Arne (1992b), Foster et al. (1991), Green (1986), Ishikawa and Tagami (1991), Issler et al. (1990), Kamp and Green (1991), Moore et al. (1986), Naeser (1979b), Omar et al. (1989), Tippett and Kamp (1993) and many others.

Many of the qualitative and quantitative properties of thermal annealing of fission tracks in apatite, which form the basis of thermal history modelling, were demonstrated in the quartet of papers by Green et al. (1986), Laslett et al. (1987), Duddy et al. (1988) and Green et al. (1989b) and in the trilogy by Carlson et al. (1999), Donelick et al. (1999) and Ketcham et al. (1999).

The book by Wagner and Van den haute (1992) gives a detailed account of the fission track dating method, while those by Dickin (2004) and Aitken (1990) contain chapters on fission track dating in relation to other dating methods in geology and archaeology.

CHAPTER 2

The Poisson line segment model

We present a simple stochastic model to describe the statistical properties of fission track measurements. The essence of this model is

(a) the locations of the fissioned uranium atoms that produce tracks form a Poisson process in three-dimensional space, and

(b) the tracks themselves are represented by line segments with random orientations and varying lengths.

It transpires that this model provides an excellent basis both for understanding fission track measurements and for quantitative analysis.

The discussion in Section 1.9 suggested that it is reasonable to assume that the locations of fissioned atoms form a Poisson process. We elaborate on this in Section 2.2. This result holds both for ^{238}U atoms that produce spontaneous tracks and for ^{235}U atoms that produce induced tracks, albeit for slightly different reasons. Furthermore, the processes that produce these two types of track are unconnected, so that, in addition,

(c) the locations of fissioned ^{238}U and ^{235}U atoms are independent.

The basic quantities that determine this stochastic process are the numbers of fissioned ^{238}U and ^{235}U atoms per unit volume and the frequency distribution of track lengths at each orientation.

Within a small enough volume of a crystal, it is reasonable to assume that the fissioned atoms are homogeneously distributed. In practice they may not be homogeneous over a larger volume — for example, if the trace uranium is more concentrated in one place than another — but for most purposes this does not matter. In particular, when counting matched pairs of spontaneous and induced tracks by the external detector method (see Chapter 3), we need to assume spatial homogeneity only very locally — within any sphere of diameter 20 microns, say. This is because both types of track emanate from essentially the same uranium source and it is just the comparison of the two matched counts that is relevant.

Furthermore, physical considerations suggest that when tracks are first formed, their lengths are independent of their orientations — that is, the length distribution is *isotropic* — and experimental observations support this. However, the thermal annealing process is not isotropic. After tracks have been partially annealed by heating, their length distribution changes with orientation.

In practice, tracks are observed because they intersect a plane surface. So a basic parameter of interest is the *track density*, usually denoted by ρ. This is the expected number of track intersections per unit area of the plane. Note that ρ is an *areal* density, in contrast to the volume density of uranium atoms that produce tracks. In the next sections we derive formulae for the expected number of tracks that intersect an arbitrary plane in terms of their lengths and orientations. Hence we obtain formulae for the track density under various conditions. In Chapter 7 we extend the argument to obtain formulae for other probability distributions and parameters.

2.1 Joint distribution of length and orientation

We start by specifying the underlying frequency distribution of track length and orientation. Let Cartesian coordinates (x, y, z) denote points in three dimensions inside the crystal, where the z-axis runs in the direction of the c-axis and the yz-plane is parallel to a prismatic face. The orientation of a track may be specified by spherical polar coordinates (θ, ϕ) defined in Figure 2.1. Here and elsewhere in this chapter, we represent three-dimensional coordinates in a two-dimensional figure, which does not always give an intuitively clear picture. It can be helpful to make model three-dimensional axes, for example, using three mutually adjacent faces of a cardboard box.

The angle between the track and the c-axis, denoted by ϕ, turns out to be informative when making inferences from track measurements. The angle θ, between the x-axis and the projection of the track into the xy-plane, is

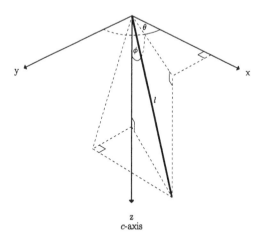

Figure 2.1 *Coordinate axes showing the track orientation angles ϕ and θ. The z-axis is parallel to the crystallographic c-axis and the yz-plane is parallel to a prismatic face. A hypothetical track of length l is drawn with one end at the origin. The track makes an angle ϕ to the z-axis (or c-axis) and its projection onto the xy-plane makes an angle θ to the x-axis.*

not informative, but is needed to define the distribution. The definition of θ strictly depends on which prismatic face the yz-plane is parallel to. However, it is natural to treat all of these faces symmetrically and hence to assume that it is immaterial which prismatic face is chosen.

To say that a line segment has random orientation means that all directions from the fissioned uranium atom have equal probability density. This implies that the joint probability density of θ and ϕ in Figure 2.1 is

$$\frac{1}{2\pi}\sin\phi, \qquad 0 \le \theta \le 2\pi, \quad 0 \le \phi \le \frac{\pi}{2}. \tag{2.1}$$

This formula may be derived as follows. Specifying an orientation is equivalent to specifying a point on the surface of a hemisphere with centre at the origin and unit radius. Random orientation means that all directions are equally probable, so all points on the surface of the hemisphere have equal probability density. Thus the probability corresponding to $d\theta\,d\phi$ (a small change in θ and ϕ) is proportional to the shaded area in Figure 2.2, which is $\sin\phi\,d\theta\,d\phi$. Normalising this so that it integrates to 1 gives the joint probability density function of θ and ϕ as in (2.1).

Note that θ and ϕ are independent and that θ has a uniform distribution but ϕ does not. In fact $\sin\phi$ has a uniform distribution between 0 and 1. It can be seen from Figure 2.2 why θ and ϕ are not both uniform: a change of $d\theta$ produces a smaller area when ϕ is small than when ϕ is large.

Let l denote the full length, in three dimensions, of an individual track and let $f(l)$ denote the relative frequency function (or probability density function)

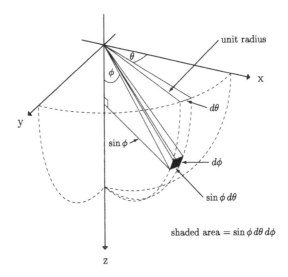

Figure 2.2 *The area on the surface of a unit sphere corresponding to a small change* $d\theta\,d\phi$ *in the polar angles* θ *and* ϕ *is* $\sin\phi\,d\theta\,d\phi$ *(after Laslett et al., 1994).*

of l. To define $f(l)$ precisely, consider a particular volume of crystal. Then $f(l)$ is the relative number of fissioned uranium atoms within this volume that give rise to tracks of length l. Part of a track may extend outside this volume, but the fissioned uranium atom must be inside. Many of the formulae in this Chapter and in Chapter 7 are derived from the idea of counting fissioned uranium atoms that produce tracks with a specific property. An important parameter is the *mean length*, which throughout this book we denote by μ, *viz.*,

$$\mu = \int_0^\infty l f(l)\, dl\,. \tag{2.2}$$

Further, let $f_\phi(l)$ denote the probability density function of lengths of tracks with a specific orientation (θ, ϕ), i.e., the conditional probability density function of length given orientation. Then $f_\phi(l)$ is the relative number of fissioned uranium atoms that produce tracks of length l, among those that produce tracks at orientation (θ, ϕ). As the notation implies, it is assumed that $f_\phi(l)$ does not depend on θ. We noted above that this distribution should not depend on which prismatic face is used to define θ — or example, for apatite it should be the same for the $\theta + k\pi/3$, where $k = 0, 1, 2, 3, 4, 5$ — and because of the physical symmetry it is reasonable to assume that it does not depend on θ at all.

The joint probability density function of length and orientation is therefore

$$f(l, \theta, \phi) = \tfrac{1}{2\pi} \sin \phi\, f_\phi(l)\,, \quad 0 \le \theta \le 2\pi\,, \quad 0 \le \phi \le \tfrac{\pi}{2}\,, \quad 0 < l < \infty. \tag{2.3}$$

The marginal probability density function of l, obtained by integration over ϕ and θ, can therefore be written as

$$f(l) = \int_0^{2\pi} \int_0^{\frac{\pi}{2}} \tfrac{1}{2\pi} \sin \phi\, f_\phi(l)\, d\phi\, d\theta = \int_0^{\frac{\pi}{2}} \sin \phi\, f_\phi(l)\, d\phi \tag{2.4}$$

and the mean track length μ can be written as

$$\begin{aligned}
\mu &= \int_0^\infty \int_0^{\frac{\pi}{2}} l \sin \phi\, f_\phi(l)\, d\phi\, dl \\
&= \int_0^{\frac{\pi}{2}} \sin \phi\, \mathrm{E}[l|\phi]\, d\phi\,,
\end{aligned} \tag{2.5}$$

where

$$\mathrm{E}[l|\phi] = \int_0^\infty l f_\phi(l)\, dl \tag{2.6}$$

is the mean length of tracks at angle ϕ to the c-axis. The function $\mathrm{E}[l|\phi]$ appears repeatedly and we use this notation throughout this book.

When the length distribution is isotropic, $f_\phi(l) = f(l)$, and in particular, $\mathrm{E}[l|\phi] = \mu$, for all ϕ. Then (2.3) simplifies to

$$f(l, \theta, \phi) = \tfrac{1}{2\pi} \sin \phi\, f(l)\,, \quad 0 \le \theta \le 2\pi\,, \quad 0 \le \phi \le \tfrac{\pi}{2}\,, \quad 0 < l < \infty. \tag{2.7}$$

2.2 The number of tracks with a given attribute

We are often interested in the number of tracks that have a specific attribute. For example, we might want to count tracks that intersect a given plane, or tracks that are longer than a given length, or both. We may think in terms of counting the uranium atoms that produce such tracks. Then it follows from the model that the number of such tracks has a Poisson distribution. This result is important in practice, because it turns out that when counting tracks it is often possible to eliminate nearly all other variation, so that the Poisson distribution can provide an intrinsic measure of observational error and a reliable basis for comparing estimates from different crystals.

A simple argument to justify the Poisson distribution appeals to the well-known limiting result for *independent thinning* of an arbitrary point process. This result says the following. Suppose that we start with an arbitrary homogeneous point process in three-dimensional space and for each point in turn, either delete it with probability $1 - p$, or retain it with probability p. Then as $p \to 0$ the locations of the *retained* points tend to form a Poisson process (e.g., Cox and Isham, 1980, §4.3). Two other properties of Poisson processes are also relevant here. Firstly, if the original points already form a Poisson process, then independent thinning will produce another Poisson process, for any p. Secondly, if two independent Poisson processes are *superimposed*, then the combined process is also a Poisson process (e.g., Cox and Isham, 1980, §4.5). Some notes on Poisson processes are given in the Appendix, Section A.1.

Now a ^{238}U atom may either decay by α-emission after a random time D or spontaneously fission after a random time S, where D is exponential with rate $\lambda_d \approx 1.55 \times 10^{-4}$ Ma^{-1}, S is exponential with rate $\lambda_f \approx 7 \times 10^{-11}$ Ma^{-1}, and S and D are independent. By a standard calculation using the generalised addition law of probability, the probability that such an atom has spontaneously fissioned by time t is therefore

$$
\begin{aligned}
p(t) &= \Pr\{S \le t, S \le D\} \\
&= (1 - e^{-\lambda_f t}) e^{-\lambda_d t} + \int_0^t (1 - e^{-\lambda_f x}) \lambda_d e^{-\lambda_d x} \, dx \\
&= \frac{\lambda_f}{\lambda_d + \lambda_f} \left(1 - e^{-(\lambda_d + \lambda_f)t}\right) \\
&= \frac{\lambda_f}{\lambda} \left(1 - e^{-\lambda t}\right),
\end{aligned}
\tag{2.8}
$$

where $\lambda = \lambda_d + \lambda_f = 1.55125 \times 10^{-4}$ Ma^{-1} is the total decay and fission rate for ^{238}U. Of course to a very close approximation $\lambda \approx \lambda_d$.

Therefore the locations of atoms that have fissioned spontaneously after a fixed time t form a thinned point process with $p = p(t)$, which is smaller than $\lambda_f / \lambda \approx 5 \times 10^{-7}$. Also, when induced tracks are created by irradiation, the probability that a particular ^{235}U atom will be induced to fission is approximately $\Phi \sigma_f$, where Φ is the applied thermal neutron fluence (in number of neutrons per square centimetre) and $\sigma_f = 580.2 \times 10^{-24}$ cm^2 is the microscopic fission cross-section of ^{235}U. The neutron fluence Φ is set by the experimenter and is practically always such that $\Phi\sigma_f$ is less than 10^{-5}.

Hence, regardless of the pattern of locations of the original uranium atoms, the locations of those that have fissioned will form a Poisson process, because each atom independently has the same very small chance of fissioning. Furthermore, this is true for spontaneous and induced fission independently.

Suppose we wish to restrict attention to tracks that have a specific attribute. We can imagine the locations of just those uranium atoms that, if they fissioned, would produce tracks with that attribute. We may apply the thinning argument just to those atoms, from which it follows that the number of such tracks will also have a Poisson distribution.

2.3 The expected number of tracks intersecting a plane

We now derive a general formula, equation (2.9), for the expected number of tracks that intersect a unit area in an arbitrary plane. Equation (2.15) gives its special form when the plane is a prismatic face. These formulae allow us to deduce the theoretical track density on various crystal faces. In Chapter 7 we generalise them by considering tracks that have additional attributes, and thence obtain formulae for other probability distributions and parameters.

The basic argument is very simple: we imagine counting the fissioned atoms within a three-dimensional volume that produce tracks with a specific attribute. Depending on the attribute, some formulae can be quite complicated when expressed explicitly in terms of the three-dimensional coordinates. Variation in three dimensions is inherent in the statistical properties of fission tracks, though, even when observing just their intersections with a plane.

Consider a crystal containing fissioned uranium atoms at concentration τ per unit volume. Suppose that fission tracks (line-segments) are counted if they intersect a specific unit area of an arbitrary fixed plane in the crystal. Let ψ be the angle of this plane to the c-axis, and without loss of generality choose the axes so that $0 \leq \psi \leq \frac{\pi}{2}$ (Figure 2.3).

Let N be the number of tracks that intersect a unit area in the plane of observation. Then from the line segment model, N has a Poisson distribution with mean, or expected value, E[N], say. We calculate E[N] in two steps: first consider just those tracks with a specific length and orientation (l, θ, ϕ) and count the number of these that intersect unit area of the plane; then add up these numbers for all (l, θ, ϕ). This argument is based on that of Parker and Cowan (1976) who give a general mathematical treatment of stochastic processes of line segments.

For definiteness, imagine that the uranium atom that produces a track is at the centre (midpoint) of the line segment. So in Figure 2.3, if the centre of a track that has length l and orientation (θ, ϕ) falls anywhere in the parallelepiped shown, then the track will intersect the unit area of surface. The expected number of such atoms is obtained by multiplying their expected number per unit volume by the volume of this parallelepiped. If we had taken

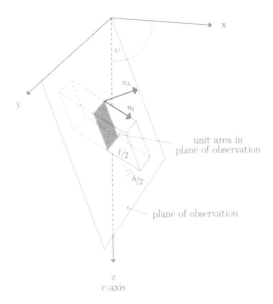

Figure 2.3 *Coordinate axes showing the angle ψ between the plane of observation and the c-axis (from Galbraith and Laslett, 1988). If the centre of a track of length l and orientation θ, ϕ falls in the parallelepiped shown, the track will intersect the unit area (shaded) in the plane of observation. The parallelepiped is of height $h = h(l, \phi, \theta, \psi)$ and the distance $h/2$, perpendicular to the plane of observation, is shown. Also shown are unit vectors n_l in the direction of the track and n_h orthogonal to the plane of observation.*

the uranium atom to be somewhere else on the line segment, the picture would be slightly different, but the relevant volume would be the same. The formula for the theoretical track density does not depend on where on the track the uranium atom is located.

The overall expected number of fissioned atoms per unit volume is τ and the relative number of these that produce tracks of length l and orientation (θ, ϕ) is given by (2.3). Hence the the expected number of atoms per unit volume that produce tracks with lengths and orientations within $(dl, d\theta, d\phi)$ of (l, θ, ϕ) is

$$\frac{\tau}{2\pi} \, d\theta \, \sin \phi \, d\phi \, f_\phi(l) \, dl \, .$$

The volume of the parallelepiped equals its base area times its height — which equals its height, since it has unit base area. Let $h = h(l, \phi, \theta, \psi)$ denote its height (Figure 2.3). Then the expected number of such tracks that intersect the unit area is

$$\frac{\tau}{2\pi} \, d\theta \, \sin \phi \, d\phi \, f_\phi(l) \, dl \, \times \, h(l, \phi, \theta, \psi) \, .$$

Adding up all of these expected numbers of tracks corresponds to integrating this product over all (l, θ, ϕ). This gives the general formula

$$E[N] = \frac{\tau}{2\pi} \int_0^{2\pi} \int_0^{\frac{\pi}{2}} \sin\phi \int_0^\infty f_\phi(l)\, h(l, \phi, \theta, \psi)\, dl\, d\phi\, d\theta. \qquad (2.9)$$

We may calculate $h(l, \phi, \theta, \psi)$ using three-dimensional geometry as follows. Let $n_1 = (0, 1, 0)$ be a unit vector in the direction of the y-axis, and let $n_2 = (\sin\psi, 0, \cos\psi)$ be a unit vector orthogonal to n_1 in the plane of observation. Then the vector product of n_1 and n_2 gives a unit vector n_h, say, orthogonal to the plane of observation (see Figure 2.3). Thus

$$n_h = n_1 \times n_2 = (\cos\psi, 0, -\sin\psi).$$

Note that the vector product $(a_1, a_2, a_3) \times (b_1, b_2, b_3)$ is defined to be the vector $(a_2 b_3 - a_3 b_2, a_3 b_1 - a_1 b_3, a_1 b_2 - a_2 b_1)$. Now let n_l denote a unit vector in the direction of a track with orientation (ϕ, θ), that is,

$$n_l = (\sin\phi\cos\theta, \sin\phi\sin\theta, \cos\phi).$$

Then the required height h is the projection of l along n_h, which is given by l times the absolute value of the scalar product of n_l and n_h. The scalar product of (a_1, a_2, a_3) and (b_1, b_2, b_3) is defined to be $(a_1 b_1 + a_2 b_2 + a_3 b_3)$. Hence

$$h(l, \phi, \theta, \psi) = l\,|n_h.n_l| = l\,|\sin\phi\cos\psi\cos\theta - \sin\psi\cos\phi|. \qquad (2.10)$$

Thus in general, $E[N]$ is given by (2.9) where $h(l, \phi, \theta, \psi)$ is given by (2.10).

2.4 Track density and equivalent isotropic length

The *track density*, denoted by ρ, is the expected number of track intersections per unit area of the observation plane. Then $\rho = E[N]$ because we are assuming that the fissioned uranium atoms are spatially homogeneous within the crystal.

The track density in an arbitrary plane is therefore obtained by substituting (2.10) into equation (2.9). After some re-arrangement, this leads to the rather complicated formula

$$\begin{aligned}
\rho = {}& \tau \sin\psi \int_0^{\frac{\pi}{2}} \sin\phi\,\cos\phi\, E[l|\phi]\, d\phi \\
& + \frac{2\tau}{\pi}\cos\psi \int_\psi^{\frac{\pi}{2}} \sin^2\phi \left\{ 1 - \frac{\tan^2\psi}{\tan^2\phi} \right\}^{\frac{1}{2}} E[l|\phi]\, d\phi \\
& - \frac{2\tau}{\pi}\sin\psi \int_\psi^{\frac{\pi}{2}} \sin^2\phi\,\cos\phi\,\cos^{-1}\left\{ \frac{\tan\psi}{\tan\phi} \right\} E[l|\phi]\, d\phi, \qquad (2.11)
\end{aligned}$$

where $E[l|\phi]$ is the mean length of tracks at angle ϕ to the c-axis, given by equation (2.6).

If the length distribution is *isotropic*, i.e., if the probability distribution of l does not depend on ϕ, then $E[l|\phi] = E[l] = \mu$, where μ is the mean track length defined by (2.2). Then the above integrals can be evaluated explicitly and (2.11) reduces to the simple formula

$$\rho = \tfrac{1}{2}\tau\mu. \tag{2.12}$$

In this case the track density is the same for any plane. Equation (2.12) is a basic formula showing how the track density is related to the mean track length for the isotropic line segment process. The corresponding formula is given in text books (e.g., Fleischer et al., 1975, Wagner and Van den haute, 1992) for the idealised case when all tracks are assumed to have the same length μ.

In general — in the isotropic or anisotropic case — we may define a parameter μ^* by the equation

$$\rho = \tfrac{1}{2}\tau\mu^*. \tag{2.13}$$

Galbraith and Laslett (1988) called μ^* the *equivalent isotropic length*, because it is the mean length of tracks that would correspond to a track density of ρ if the length distribution was isotropic. This quantity is a basic parameter for the anisotropic line segment process and it generally plays the same role there that the mean length μ plays in the isotropic process.

2.5 Tracks intersecting a prismatic face

An important special case is when the observation plane is a *prismatic face*, so that $\psi = 0$. This is the usual situation for observing tracks in an apatite crystal. Equation (2.10) then simplifies to

$$h(l,\phi,\theta,0) \;=\; l\sin\phi\,|\cos\theta|. \tag{2.14}$$

Substituting equation (2.14) in (2.9), the outer integral, over $0 \le \theta \le 2\pi$, is just four times the integral over $0 \le \theta \le \tfrac{\pi}{2}$. Hence, for a prismatic face, equation (2.9) simplifies to

$$\mathrm{E}[N] \;=\; \frac{2\tau}{\pi}\int_0^{\frac{\pi}{2}}\cos\theta\int_0^{\frac{\pi}{2}}\sin^2\phi\int_0^{\infty} l\,f_\phi(l)\,dl\,d\phi\,d\theta. \tag{2.15}$$

Furthermore, integrating over θ and using (2.6) this may be written as

$$\rho \;=\; \frac{2\tau}{\pi}\int_0^{\frac{\pi}{2}}\sin^2\phi\,\mathrm{E}[l|\phi]\,d\phi, \tag{2.16}$$

a formula originally derived by Laslett et al. (1984). This formula may also be obtained by putting $\psi = 0$ in (2.11).

For tracks intersecting a prismatic face, the equivalent isotropic length is thus given by

$$\mu^* \;=\; \frac{4}{\pi}\int_0^{\frac{\pi}{2}}\sin^2\phi\,\mathrm{E}[l|\phi]\,d\phi. \tag{2.17}$$

2.6 Effect of non-prismatic face on track density

To illustrate some numerical consequences of the above formulae, let us see how the track density might depend on the angle ψ between the plane of observation and a prismatic face.

In observing and counting fission tracks in apatite, practitioners have advocated that the plane of observation should be a prismatic face, because then the etching efficiency is highest, and close to 100% (Gleadow, 1978). That is, practically all tracks that intersect a prismatic face will be etched, whereas this is not so for a basal face, or for one at a high angle to a prismatic face. Furthermore, if the observation plane is assumed to be prismatic, but in fact is not so, then the true track density may be under-estimated.

But, as the above theory suggests, there is also a potential bias in track counts arising from geometrical considerations. If the length distribution is isotropic, then the true track density ρ of tracks intersecting any plane is the same. But if the length distribution is anisotropic, then ρ does depend on the orientation of the plane of observation, quite apart from any considerations of etching efficiency. In practice is is difficult to ensure that the plane of observation is exactly parallel to a prismatic face, so that variation in the angle ψ could in theory contribute to variation in fission track age estimates from different crystal grains in annealed samples.

We may calculate how ρ depends on the angle ψ using equation (2.11). In Chapter 7, Section 7.8, we note that the integrals in this equation may be evaluated explicitly if we approximate $E(l|\phi)$ by a polynomial in $\cos\phi$, and we illustrate the calculation using a quadratic polynomial with coefficients

Table 2.1 *Variation of track density (equivalent isotropic length in μm) with angle ψ between the observation plane and a prismatic face, and with amount of annealing (from Galbraith and Laslett, 1988).*

Angle ψ	Equivalent isotropic length from equation (7.42)				
	Unannealed	260°C	310°C	336°C	352°C
0°	16.41	14.67	12.81	10.03	5.75
$7\frac{1}{2}°$	16.42	14.67	12.82	10.05	5.79
15°	16.43	14.69	12.84	10.08	5.89
$22\frac{1}{2}°$	16.46	14.71	12.87	10.13	6.06
45°	16.55	14.81	13.03	10.36	6.96
90°	16.62	14.93	13.24	10.64	8.31
Percent relative increase compared with $\psi = 0$					
$7\frac{1}{2}°$	.03	.04	.06	.12	.61
15°	.12	.14	.23	.46	2.40
$22\frac{1}{2}°$	.25	.31	.51	.99	5.28
45°	.85	.95	1.72	3.29	21.04
90°	1.28	1.77	3.36	6.08	44.52

estimated empirically from observed lengths and angles of confined tracks. Table 2.1 is based on this method of calculation and gives us a feel for the likely amount of such variation in track density that may be due to variation in ψ. In this table the equivalent isotropic length μ^* in microns is used as a surrogate for the track density ρ. The two are proportional of course, but the former can be calculated from lengths and angles alone.

Table 2.1 shows how μ^* may vary with ψ for various amounts of annealing. As ψ increases, track density increases, and more so for more heavily annealed samples, because of the increasing anisotropy. In general, the relative increase (compared with $\psi = 0$) is small even for quite large angles ψ, but the differences are noticeable at high angles for the heavily annealed samples.

It would be hard to study this effect experimentally because of difficulty in measuring ψ, but it seems likely from these calculations that, provided ψ is small, loss of etching efficiency is a more important consideration than the effect of anisotropy of the track length distribution.

2.7 Track counts from a dosimeter glass

As a small empirical illustration of Poisson variation, consider the counts in Figure 2.4, which were obtained from fission tracks induced in a *dosimeter glass*. This is an artificially produced glass designed to contain a known amount of uranium, distributed uniformly throughout. The glass was clamped to the mica detector and irradiated in a nuclear reactor. Induced fission tracks near the glass surface may emerge and create tracks in the mica which can be observed. Dosimeter glasses are used to measure the neutron fluence in an irradiation by calibrating the number of tracks produced against the known amount of uranium. The counts in Figure 2.4 are the numbers of fission tracks intersecting the detector in each of 64 equally sized square areas (fields) arranged in an 8 × 8 grid, according to the figure.

If the uranium concentration in the glass is spatially uniform, the expected count should be the same in each field, and therefore the counts when taken

```
 95  90  86  81  96  89  78  73
 77  90  79  92  84  87  82  81
 86  90  71  96 107  84 100  73
 92  92  99 105  82  98  96  78
 91  79 109  79  80  87  83  71
 93  72  90  80  64  85  71  90
102 103  84 103  96  88  84  83
 90  90  94  91  71  74  85  88
```

Figure 2.4 *Spatial distribution of track counts 64 fields in an 8 × 8 grid in CN5 glass. The area of one field is 69.25 × 10^{-6} cm^2. These data were kindly supplied by Andrew Carter from the UCL Fission Track Laboratory.*

together should vary according to a Poisson distribution. There are various ways to assess whether a set of n observed counts might have arisen from a Poisson distribution. A useful one here is to calculate the Poisson *index of dispersion*, which is defined as $(n-1)$ times the sample variance divided by the sample mean count. If the counts are from a Poisson distribution, this index should be consistent with a value from a χ^2 distribution with $(n-1)$ degrees of freedom. A significantly large index would indicate that the counts vary more than they should, possibly in this case because the uranium is heterogeneous. A significantly small index would indicate that the counts vary *less* than they should — i.e., that they are *under-dispersed* with respect to Poisson variation.

The mean and variance of the 64 counts in Figure 2.4 are, respectively, 86.86 and 97.11. The Poisson index of dispersion is thus $63 \times 97.11/86.86 = 70.43$, with 63 degrees of freedom. This is consistent with Poisson variation, the upper $\chi^2(63)$ tail probability (i.e., the p-value) being 0.24.

A more convincing analysis is to test for departure from Poisson variation for different equal-sized areas. For example, it is possible that counts in small areas may vary consistently with a Poisson distribution while counts pooled into larger equal areas do not. But the model implies that Poisson variation should hold regardless of the size of the fields of counts. An efficient method of doing this was devised by Greig-Smith (1952) and is described, for example, by Ripley (1981, §6.3). In our example (Figure 2.4), we compare the two total counts for the top and bottom halves, then the four counts for quarters within halves, then the eight counts for eighths within quarters, and so on. At each level a χ^2-statistic is calculated, representing a Poisson index of dispersion for that level of pooling. Table 2.2 shows the results. None of the p-values is significantly small and the variation is clearly consistent with Poisson variation at every level.

Table 2.2 *Poisson dispersion analysis of the data in Figure 2.4*

Number of fields	Level of variation	χ^2_{stat}	d.f.	p
32	4×8 within 8×8	1.44	1	.23
16	4×4 within 4×8	0.73	2	.69
8	2×4 within 4×4	5.94	4	.20
4	2×2 within 2×4	5.08	8	.75
2	1×2 within 2×2	8.40	16	.94
1	1×1 within 1×2	37.49	32	.23
Overall	1×1 within 8×8	70.43	63	.24

χ^2_{stat} denotes the Poisson index of dispersion at the relevant level, d.f. denotes the degrees of freedom and p is the upper χ^2 tail probability.

2.8 Spatial and temporal variation

Most fission track counts from standard uniform sources behave similarly to those in Figure 2.4. Occasionally there is evidence of over-dispersion — one or more of the χ^2-statistics being too large — but this is more plausibly regarded as evidence of spatial heterogeneity in uranium concentration, or of some other experimental factor, rather than evidence against the validity of the line segment model.

As implied above, there might be evidence of *under*-dispersion with respect to Poisson variation. For fission track counts it is implausible that this could arise naturally. In point processes, sub-Poisson variation can arise from inhibition between points (e.g., where an atom fissioning inhibits other atoms nearby from fissioning) or by non-random selection of points (e.g., where some tracks are not counted if they are near other tracks). Neither of these phenomena should apply here.

Spontaneous and induced track counts in apatite and other minerals also exhibit Poisson variation. This can sometimes be seen directly, as above, when the underlying uranium concentration is spatially uniform. But usually the uranium concentration is not spatially uniform. Then counts from different areas will exhibit *over*-dispersion with respect to Poisson variation. This of course does not invalidate the model but merely reflects the fact that each count may be from a Poisson distribution with a different mean — i.e., the counts are over-dispersed because their Poisson means vary. This is a phenomenon that has been misunderstood in the past.

For the external detector method of fission track dating, which we deal with in Chapter 3, the component of dispersion that is due to spatial variation in uranium is eliminated. This is because for each count of spontaneous tracks there is a matched count of induced tracks generated from the same uranium source, albeit from a different isotope. For such matched pairs of counts from different grains of the same crystal, or from different crystals of the same age, the expected counts may vary but they will be in a fixed ratio. It then transpires that the *residual* variation is again consistent with the Poisson model. This has been seen so often in practice that any departure from Poisson variation in this sense is usually a sign that it is due to some other cause, rather than evidence against the Poisson model as such.

We have emphasised the derivation of the Poisson distribution for track counts by way of the underlying Poisson process of fissioned uranium atoms in three-dimensional space. Others have argued that the counts have Poisson distributions because fissions take place randomly in time — that is, in a *temporal* Poisson process. While this is true, it is not directly relevant to the way tracks are observed. Spontaneous tracks are observed spatially and at a single time, not sequentially at different times. The temporal process does not say anything about spatial variation. Nor do counts of tracks, by themselves, tell us anything about the time order in which they were formed, though information about this can sometimes be inferred from track length measurements if the mineral has been heated. For induced tracks, the temporal process of fissioning is a less natural concept than the spatial one and again does not describe the detailed observations in the crystal.

2.9 Remarks

Like all mathematical models, the Poisson line segment model is an idealisation. But it does reflect the essential features of observed data very well. When apparent departures from the model are seen, there is nearly always an identifiable reason — for example, factors such as uranium concentration, temperature, age or chemical composition may be varying — so that the model then provides a sound basis for studying these factors. This model is therefore important both theoretically, to enable us to understand how different measurements may vary, and practically, to provide reliable estimates and their precisions — and to build more specialised models.

In fact, to justify specifically the Poisson distributions of track counts, it is not strictly necessary that tracks be represented as line segments. It would be sufficient that any track produced by an atom at a given distance from the plane had the same chance of intersecting the plane. But the line segment model provides a deeper understanding of the role of track lengths in relation to track densities and other parameters.

2.10 Bibliographic notes

Parts of this chapter follow Galbraith and Laslett (1988). Stochastic processes of random line segments arise in several other contexts, including random fibrous networks, random drift sampling and particle tracks through polymer solutions. They have been studied by a number of authors, including Coleman (1972, 1974), Kallmes *et al.* (1961), Lehman and Brisbane (1968), Ogston (1958) and Ogston *et al.* (1973). A general treatment is given by Parker and Cowan (1976).

Some formulae derived here and in Chapter 7 can be expressed more concisely in terms of solid angle representations. For example, random orientation means that the solid angle is uniform. But by using the polar angles θ and ϕ we can obtain explicit formulae in terms of the important angle ϕ.

CHAPTER 3

Track counts and densities: fission track dating

Fission track dating is based on the idea that, because tracks form continuously over geological time, the number of spontaneous fission tracks counted per unit area in a crystal will indicate its age. Of course this number depends on other factors too, notably the amount of uranium the crystal contains. This latter is usually estimated indirectly by creating induced fission tracks in the same sample and counting them. There are other calibration requirements and procedures that are necessary before one can sensibly estimate a fission track age. This chapter is about both the underlying mathematical theory and the statistical analysis of fission track counts in this context.

Why is statistical analysis necessary? Observed fission track data are the result of various natural and experimental processes, all of which are subject to variation. Statistical analysis is needed to understand the data in the presence of this variation. We present some simple methods designed to aid such understanding. These may be applied routinely, although each sample should be placed in its geological context, and further analysis specific to that context may be needed.

The interpretation of fission track data is not straightforward. A newcomer to the field may have some preconceptions, such as

- all grains from a single sample have the same age,
- the time over which tracks have been accumulating in each grain is the same as the fission track age of the sample, or
- data from each grain may be pooled to estimate a common age for the sample.

Departures from these assumptions occur so often that it would be dangerous to entertain them except as initial working hypotheses. Indeed much of the present day interest in fission track analysis is because it is more than just a dating technique. Thus, grains from a single sedimentary sample may have been transported from more than one source and hence may have different ages of formation. The retention of fission tracks is temperature dependent, particularly in apatite, and tracks will not accumulate in any significant way until the temperature has dropped below a certain level. So the fission track age, if it has a meaning in its own right, is a cooling age. Furthermore, the thermal behaviour of fission tracks in apatite depends upon its chemical composition, so that if the chemical compositions vary, each grain in a sample may have a different cooling age.

Despite this, the pooling of data across grains to calculate the *pooled fission track age* is a common *de facto* protocol, and one objective of this chapter is to present a more realistic protocol for the analysis of fission track data. Central to this approach is the notion of a *true fission track age* for each grain in a sample. This is the time t that would be calculated from the fission track age equation if every quantity in that equation could be measured with complete precision — that is, if sufficient high-quality data were available. Now, the true fission track age does not necessarily correspond to the time of any real event, because of the effect of heat upon fission tracks, so that the true fission track age reflects not only the age of formation but also the thermal history the grain has experienced and its chemical composition. The *true* fission track ages may therefore differ from grain to grain. In addition, *estimates* of the fission track age will differ from grain to grain in a sample even when the true value is the same for each grain.

This chapter gives both graphical and formal statistical methods for detecting when the true fission track ages may conform to a single value. A second consideration, common to all statistical problems, is that only a sample of data is available, so that the true fission track age cannot be calculated exactly, but can only be estimated. Another objective of this chapter is to provide a routine protocol for calculating single grain age estimates and their precisions.

In the next section we derive the mathematical basis for fission track dating, based on the line segment model of Chapter 2. We then develop the theory for the external detector method. This is the most popular and potentially most useful experimental method, mainly because variation in uranium concentration between (and to some extent within) grains, which is often considerable, is eliminated as a source of error variation. In Chapter 4 we present corresponding methods for data obtained by an alternative procedure, called the *population method*, and we discuss briefly a variant of it, the *population subtraction method.*

3.1 The mathematical basis of fission track dating

Originally, Price and Walker (1963) derived their fission track age equation, equation (3.6) below, using a deterministic argument. Here we give a stochastic argument in order both to establish the age equation and to highlight the statistical assumptions associated with it.

Consider a crystal (e.g., apatite or zircon or sphene) containing trace uranium with ^{238}U atoms that have been undergoing spontaneous fission for a geological time t, where t is usually measured in millions of years. As noted in Chapter 2, a ^{238}U atom may either decay by α-emission after a random time D or spontaneously fission after a random time S, where D and S are independent exponential random variables with rates $\lambda_d \approx 1.55 \times 10^{-4}$ Ma^{-1} and $\lambda_f \approx 7 \times 10^{-11}$ Ma^{-1}, respectively. The probability that such an atom has spontaneously fissioned by time t is, from equation (2.8),

$$p(t) = \frac{\lambda_f}{\lambda}\left(1 - e^{-\lambda t}\right),$$

where $\lambda = \lambda_d + \lambda_f = 1.55125 \times 10^{-4}$ Ma^{-1} is the total decay and fission rate for ^{238}U.

Let τ_0 denote the concentration of ^{238}U atoms (that is, the expected number per unit volume) at the time of formation of the host crystal. And let τ_s denote the concentration of spontaneously fissioned ^{238}U atoms at time t. Then

$$\tau_s = \tau_0 \, p(t) = \tau_0 \frac{\lambda_f}{\lambda}\left(1 - e^{-\lambda t}\right).$$

Also, if τ_c is the current (present day) concentration of ^{238}U atoms, then

$$\tau_c = \tau_0 \Pr\{S > t, D > t\} = \tau_0 \, e^{-\lambda_f t} e^{-\lambda_d t} = \tau_0 \, e^{-\lambda t}$$

so that

$$\tau_s = \frac{\lambda_f}{\lambda}\left(e^{\lambda t} - 1\right)\tau_c. \tag{3.1}$$

Hence, if the ratio τ_s/τ_c could be measured, the age t of the crystal could be inferred.

Information about τ_s is obtained by counting tracks that intersect a polished planar surface within the body of the crystal. These tracks are formed prior to polishing, and arise from fissioned ^{238}U atoms on both sides of the surface. Under the line segment model, the expected number of such tracks (i.e., intersections with the plane) per unit area is given by equation (2.13), which we now write as

$$\rho_s = \tfrac{1}{2}\tau_s \mu_s = \frac{\lambda_f}{2\lambda}\left(e^{\lambda t} - 1\right)\tau_c \, \mu_s, \tag{3.2}$$

where ρ_s is called the *spontaneous track density* and μ_s is the mean length (or more correctly, the *equivalent isotropic length*) of spontaneous tracks. When λt is small, which is the usual case, $(e^{\lambda t} - 1)/\lambda \approx t$ so that

$$\rho_s \approx \tfrac{1}{2}\lambda_f \tau_c \mu_s \, t.$$

Although there is no need to make this approximation in calculations, it is helpful conceptually to remember that ρ_s is essentially proportional to t.

Information about τ_c is obtained indirectly, effectively by measuring the current concentration of the ^{235}U atoms, τ_{235} say, and using the fact that the current ^{235}U:^{238}U isotopic ratio is known. It is usually denoted by I and its value is $I = \tau_{235}/\tau_c = 7.25 \times 10^{-3}$ (see Section 1.9).

Measurement of τ_{235} is done by irradiating crystals with thermal neutrons, thus inducing a measurable proportion of ^{235}U atoms to fission, and creating *induced* fission tracks. The probability that a ^{235}U atom is induced to fission is $\Phi\sigma_f$, where Φ is the applied thermal neutron fluence, in number of neutrons per square centimetre, and $\sigma_f = 580.2 \times 10^{-24}$ cm^2 is the microscopic fission cross-section of ^{235}U. The neutron fluence Φ is set by the experimenter and is practically always such that $\Phi\sigma_f$ is less than 10^{-5}.

Let τ_i be the expected number of *fissioned* ^{235}U atoms per unit volume. Then

$$\tau_i = \Phi\sigma_f\tau_{235} = \Phi\sigma_f I\tau_c$$

and hence

$$\frac{\tau_s}{\tau_i} = \frac{\lambda_f}{\lambda\Phi\sigma_f I}\left(1 - e^{-\lambda t}\right). \tag{3.3}$$

Various experimental methods have been proposed to estimate the ratio τ_s/τ_i and hence to estimate t, the other quantities in equation (3.3) being known or measurable. These include the population method, the re-etch method, the external surface method, and many variants. But the most popular and most useful method, particularly for the dating of apatite, zircon and sphene, is the *external detector method*, which we describe in more detail in the next section.

In this method, induced tracks intersecting the same polished area of plane surface as used for the spontaneous tracks are recorded in a mica detector and counted. The expected number of induced tracks per unit area intersecting this plane, called the *induced track density* ρ_i, is given by

$$\rho_i = \tfrac{1}{4}\tau_i\mu_i = \tfrac{1}{4}\Phi\sigma_f I\tau_c\mu_i, \tag{3.4}$$

where μ_i is the mean length of induced tracks. The factor $\frac{1}{2}$ in (3.2) becomes $\frac{1}{4}$ here, because the induced tracks arise from ^{235}U atoms on only one side of the crystal surface. That is, the relevant volume containing ^{235}U atoms is half the size of that used for the ^{238}U atoms.

Using equations (3.2) and (3.4) to substitute for τ_s and τ_i in terms of ρ_s and ρ_i in equation (3.3), and rearranging, leads to

$$t = \frac{1}{\lambda}\log\left(1 + \frac{1}{2}\frac{\lambda}{\lambda_f}I\,\sigma_f\,\Phi\,\frac{\rho_s}{\rho_i}\frac{\mu_i}{\mu_s}\right). \tag{3.5}$$

This is analogous to the equation originally derived by Price and Walker (1963) using a deterministic argument — that is, by arguing in terms of mean values — which is the basis of most derivations in the literature. The present argument, using the Poisson line segment model, highlights the relevance of μ_s and μ_i and also allows us to infer appropriate statistical distributions for the track counts. Price and Walker implicitly assumed that the track length distribution is isotropic and that $\mu_s = \mu_i$ in order to obtain their *fission track age equation* for the time t over which tracks have been forming, *viz.*,

$$t = \frac{1}{\lambda}\log\left(1 + \frac{\lambda I\sigma_f\Phi}{2\lambda_f}\frac{\rho_s}{\rho_i}\right), \tag{3.6}$$

where

$\lambda = 1.55125 \times 10^{-4}\text{Ma}^{-1}$ is the total decay rate of ^{238}U,
$\lambda_f \approx 7 \times 10^{-11}\text{Ma}^{-1}$, is the spontaneous fission rate of ^{238}U,
$I = 7.25 \times 10^{-3}$ is the ^{235}U:^{238}U isotopic ratio,

$\sigma_f = 580.2 \times 10^{-24}$ cm^2 is the cross-section for induced fission of ^{235}U,

Φ is the applied thermal neutron fluence in cm^{-2}

and ρ_s and ρ_i are the areal densities of spontaneous and induced fission tracks, usually measured in millions of tracks per square centimetre. With these units, equation (3.6) gives t in Ma. This equation shows that the fission track age is a function of the *ratio* of the spontaneous and induced track densities. Once the neutron fluence Φ is determined, estimating the fission track age is tantamount to estimating the ratio ρ_s/ρ_i.

Note, however, that if tracks have undergone thermal annealing, then μ_s will be less than μ_i for two reasons. Firstly, spontaneous tracks will have shortened and secondly, their length distribution will not be isotropic so that the equivalent isotropic length will be less than the mean length. In this case, equation (3.6) will under-estimate the time over which tracks have been forming.

3.2 The external detector method

This experimental design is used routinely because of its practical convenience, high precision, and single grain dating capability (e.g., Gleadow, 1981). Furthermore, its statistical basis is well established and requires very few arbitrary assumptions. It is used for the common minerals apatite, zircon and sphene.

3.2.1 Experimental method

After separation from the host rock, the mineral grains are mounted in epoxy resin or Teflon®. The mount is ground and polished with aluminium oxide paste to expose flat interior surfaces of several grains and is then etched (for apatite, this might be with 5 Molar HNO$_3$ for 20 seconds at 20°C). This reveals spontaneous tracks that intersect these surfaces. An external detector containing negligible uranium, commonly a flat piece of mica, is then attached to the mount and the mount and detector are irradiated together. Tracks from induced fission of ^{235}U atoms near the polished surfaces may emerge from inside the grains and create regions of damage within the mica. After irradiation, the mica is removed from the mount and is etched to reveal these induced tracks. Figure 3.1 shows a photograph of some etched spontaneous tracks in apatite and some induced tracks in the matched area of a mica detector.

Counts of tracks are made by a trained observer using an optical microscope linked to a computerised recording system. The observer decides which grains, or part grains, are suitable for counting tracks in, according to laboratory practice. For each, a rectangular grid of graticule squares is placed over the grain's surface. The number of spontaneous fission tracks intersecting the surface in each graticule square of the grid is counted, according to well-defined observational criteria and with care to avoid edge effects, for example with parts of the grain that contain cracks. The number of graticule squares over which tracks are counted is also recorded and is later used to

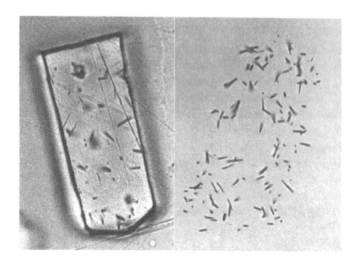

Figure 3.1 *The external detector method. The left panel shows etched spontaneous tracks intersecting a polished surface in an apatite crystal. The right panel shows etched induced tracks intersecting the matched area of a mica detector. Photograph courtesy of Paul Green.*

determine the corresponding area of surface. The observer places a similar grid on the corresponding part of the mica detector, and then counts induced fission tracks in the matched squares in the mica.

The observer needs to be sure that the same parts of mineral grain and detector are both suitable for counting. This is normally determined by the quality of the mineral grain as tracks in the mica detector are much easier to identify and count. The density of tracks in the detector depends not only on the amount of uranium in the mineral but also on the neutron fluence used in the irradiation, which can be controlled by the experimenter. When choosing the neutron fluence, the experimenter wants to produce enough tracks to get a good measure of ρ_i but not so many that they are too dense to count.

3.2.2 Some typical data

Table 3.1 presents fission track counts and counting areas from 20 apatite grains separated from a rock sample with an age of 162 Ma, independently determined by ^{40}Ar/^{39}Ar dating. The term "grain" is generally used to mean a single crystal or part of crystal on which a separate count is made. The sample was used as an age standard in an inter-laboratory comparison (Miller *et al.*, 1993). The data were generated by an anonymous analyst using the external detector method, in accordance with standard practice. For example, grain 1 has 31 spontaneous tracks and 41 induced tracks counted over an area of 40

Table 3.1 *Spontaneous and induced track counts obtained by the external detector method for 20 apatite grains (sample 92-1, Miller et al., 1993).*

Grain	N_s	N_i	Area	Grain	N_s	N_i	Area
1	31	41	40	11	35	52	40
2	19	22	20	12	52	76	70
3	56	63	60	13	51	74	49
4	67	71	80	14	47	66	50
5	88	90	90	15	27	39	36
6	6	7	15	16	36	44	40
7	18	14	20	17	64	86	50
8	40	41	40	18	68	90	50
9	36	49	40	19	61	91	60
10	54	79	60	20	30	41	30

N_s = number of spontaneous tracks counted; N_i = number of induced tracks counted; area = number of graticule squares counted (area of one graticule square = 0.9009 $\times$ 10^{-6} cm^2); $\hat{\zeta}$ = 350 Ma $\times 10^{-6}$ cm^2; relative standard error of $\hat{\zeta}$ = 10/350 = 0.029; N_d = 2936 tracks; $\hat{\rho}_d$ = 1.304 $\times$ 10^6 cm^{-2}.

graticule squares. The area of a graticule square is determined by calibration and in this instance is 0.9009×10^{-6} cm^2.

Table 3.1 is typical of a set of homogeneous fission track counts. The ratios of spontaneous to induced counts vary between 0.67 and 1.29, with a median of 0.76. This variation is quite small and might be consistent with the hypothesis that the ratio of true track densities is the same for all grains, which would imply that they have a common age. Part of the aim of a statistical analysis is to assess whether this is the case and, if so, to estimate the common age. The areas in Table 3.1 also vary, because apatite grains come in a range of sizes, and because the exposed surface of a grain is determined by random plane sampling. However, the variation of the areas in Table 3.1 is modest compared with that in many field samples.

It is usual to quote, as part of the data, measured values of two laboratory calibration constants, ζ and ρ_d. Here $\hat{\zeta}$ = 350 Ma $\times 10^{-6}$ cm^2 and $\hat{\rho}_d$ = 1.304 $\times 10^6$ tracks per square cm. The ζ value is characteristic of each analyst, and ρ_d is an indirect measure of the thermal neutron fluence used to irradiate the sample. The conventional units of ζ and ρ_d are such that the product $\zeta \rho_d$ is in Ma. More detailed explanation of ζ and ρ_d is given in Section 3.5. The "hat" notation, as in $\hat{\zeta}$ and $\hat{\rho}_d$, indicates that these quantities are estimated empirically and will have associated numerical standard errors. Knowing these constants, a fission track age may be calculated for each grain and, if appropriate, a common age may be calculated for the sample, using equation (3.12) below.

Table 3.2 presents track counts and areas for 30 apatite grains from a granite from the Seychelles. These data were kindly provided by Andrew Gleadow. By

Table 3.2 *Spontaneous and induced track counts obtained by the external detector method for 30 apatite grains from the Mahe granite, Seychelles. Data provided by Andrew Gleadow.*

Grain	N_s	N_i	Area	Grain	N_s	N_i	Area
1	0	11	24	16	10	17	16
2	2	11	44	17	2	5	12
3	18	28	32	18	7	23	40
4	2	4	40	19	1	10	60
5	10	78	90	20	14	43	24
6	3	22	30	21	15	44	49
7	4	8	40	22	14	25	48
8	20	57	50	23	8	28	32
9	52	129	20	24	22	69	45
10	2	7	45	25	16	29	44
11	1	9	35	26	34	51	35
12	6	16	50	27	14	56	40
13	256	220	100	28	6	9	5
14	52	134	24	29	13	22	32
15	3	11	35	30	127	213	28

N_s = number of spontaneous tracks counted; N_i = number of induced tracks counted; area = number of graticule squares counted (area of one graticule square = 0.78 $\times$ 10^{-6} cm^2); $\hat{\zeta}$ = 380 Ma $\times 10^{-6}$ cm^2; relative standard error of $\hat{\zeta}$ = 5/380 = 0.013; N_d = 8188 tracks; $\hat{\rho}_d$ = 1.257 $\times$ 10^6 tracks cm^{-2}.

comparison with Table 3.1, the spontaneous track counts are more variable, ranging from 0 to 256, and the ratios of spontaneous to induced track counts range more widely, from 0 to 1.16. An experienced analyst would infer, for example, that grain 5, with N_s = 10 and N_i = 78, is younger than grain 13, with N_s = 256 and N_i = 220. In some grains in this sample, tracks were counted over small areas, such as grain 17 (with 12 graticule squares) and grain 28 (5 squares). Here the area of a single graticule square is 0.78 $\times$ 10^{-6} cm^2. Note also that grain 30 has large track counts over a relatively small area, indicating a high uranium concentration. Statistical methods of analysis must be able to cope with such variation in counts and areas.

It is quite common for the ratios N_s/N_i to vary even more widely than in Table 3.2, for example from 0 to 3 in the same sample. The analyst needs to understand such variation. This would normally involve using suitable graphical displays, numerical summaries and diagnostics, and perhaps fitting models that reflect the geological environment.

3.3 Observed and theoretical track densities

The line segment model implies that, for a single grain dated by the external detector method the numbers of spontaneous and induced tracks, N_s and

N_i, counted over matched areas A have independent Poisson distributions with means $A\rho_s$ and $A\rho_i$, where ρ_s and ρ_i are the *theoretical* spontaneous and induced track densities. These are important quantities, which we now explain briefly.

In the fission track literature it is common to denote the *observed* track densities by ρ_s and ρ_i. In this book it is necessary to use the more rigorous statistical notation, whereby the observed track densities are $\hat{\rho}_s$ and $\hat{\rho}_i$, and the true densities are denoted by ρ_s and ρ_i. Consider hypothetical replicates of a single grain, i.e., with the same uranium content, surface area and environment. Then ρ_s represents the average number of spontaneous tracks per unit area, averaged over all hypothetical replicates. In practice we observe just one count N_s for that grain out of a conceptual infinite number, so that N_s may be regarded as an observed value from a Poisson distribution with mean $A\rho_s$. This mean is often called the *expected* count, though it need not be an integer of course. The observed density $\hat{\rho}_s$ is equal to N_s/A, and is an *estimate* of ρ_s. A similar interpretation applies to ρ_i and $\hat{\rho}_i$.

Moreover, the Poisson distribution with mean $A\rho_s$ has standard deviation $\sqrt{A\rho_s}$, representing the standard deviation of all conceivable counts that might arise in the hypothetical replications. If we divided each hypothetical count by A we can imagine an infinite set of *estimates* of ρ_s, and the standard deviation of these is $\sqrt{\rho_s/A}$. This is the *standard error* of $\hat{\rho}_s$ — the standard deviation of the possible estimates that "might have been obtained" in the experiment. Again, a similar interpretation applies to $\hat{\rho}_i$ and indeed to other quantities that are estimated from experiments. Some notes on standard errors are given in the Appendix, Section A.4.

Now a property of the Poisson distribution is that, if two or more independent counts are added then the total count also has a Poisson distribution, with mean equal to the sum of the individual means. For example, $N_s + N_i$ has a Poisson distribution with mean $A\rho_s + A\rho_i$. Furthermore, if we add the counts of spontaneous tracks over several grains, and similarly add the counts of induced tracks, then the two total counts will be a matched pair of Poisson counts in just the same way as for a single grain. This implies that the same standard error formula also applies to a track density estimated from a sum of counts, where A is the total area over which tracks were counted and ρ_s is the total expected number of tracks divided by A. We simply regard the two total counts as a pair of matched counts from a "super-grain" consisting of all of the individual grains.

In order to calculate the standard error, we need the value of ρ_s, which of course we do not know. It is common practice to substitute the estimate $\hat{\rho}_s$ to obtain the approximate standard error $\sqrt{\hat{\rho}_s/A} = \sqrt{N_s}/A$. Another common practice is to quote a *relative* standard error, which is the standard error divided by the true value of the parameter — in this case

$$\frac{\text{se}(\hat{\rho}_s)}{\rho_s} = \sqrt{\frac{1}{A\rho_s}} \approx \sqrt{\frac{1}{N_s}}.$$

Similarly, for the estimate of ρ_i,

$$\frac{se(\hat{\rho}_i)}{\rho_i} = \sqrt{\frac{1}{A\rho_i}} \approx \sqrt{\frac{1}{N_i}}.$$

3.4 A short digression

There is a hitherto unmentioned feature of data obtained by the external detector method that has caused some confusion in the past. Looking at the pairs of counts in Table 3.1, one can see that they tend to go up and down together. For example, for grain 5 they are both high and for grain 6 they are both low. In other words, they are *correlated*. If we plotted a graph of the N_s values against the N_i values they would scatter quite closely about a line — in fact, about a line through the origin. How does this square with our assertion that N_s and N_i are independent?

Of course, N_s and N_i vary together because they are counts over the same area and from the same uranium source, and both the areas and the uranium contents vary between grains. Although N_s and N_i are independent counts from Poisson distributions with means $A\rho_s$ and $A\rho_i$, these means vary between grains. In one sense, therefore, they are correlated, but this does not invalidate the above standard error formulae, as the "super-grain" interpretation shows. In fact it is unhelpful to think of this phenomenon in terms of *correlation*. It is best understood as *variation* — i.e., variation in uranium content between grains — and furthermore, variation that is immaterial to the estimation of ρ_s/ρ_i.

The concept of a standard error of an estimate is quite subtle and is discussed in the Appendix, Section A.4, but there is a further aspect that merits mention here. The "hypothetical replicates" that we referred to really are hypothetical. We can imagine them but it is usually not possible to do them, for example because we could not obtain grains, or even parts of grains, with the same uranium contents and other features. Now, one method of mimicking replicates is to do computer simulations. And in doing this, we need to think about which parameters should be fixed for all replicates and which should be allowed to vary. That is, exactly what constitutes a repeat measurement? This question has more than one answer in general. In effect, different answers may address different aspects of the experimental process.

In the present case, we may imagine giving fixed values to ρ_s, ρ_i and A and generating a random value from a Poisson distribution with mean $A\rho_s$ and one from a Poisson distribution with mean $A\rho_i$ to get hypothetical values of N_s and N_i. If we repeated this many times, to get many pairs of counts, and many estimates of ρ_s and ρ_i, these estimates would indeed vary with the standard deviations given above, viz., $\sqrt{\rho_s/A}$ and $\sqrt{\rho_i/A}$, respectively. But we can imagine other methods of simulation also. We could fix ρ_s and ρ_i but let A vary between replicates. What difference would this make?

More subtly, we could let ρ_s, ρ_i and A all vary but keep the *ratio* ρ_s/ρ_i fixed. If we simulated 20 pairs of counts in this way we might get data that looked something like those in Table 3.1. Then if we looked at the separate estimates of ρ_s and ρ_i, they would *not* vary with the above standard deviations because they are estimates of different quantities. But a surprising thing happens. If for each pair of counts we calculated their *ratio*, which is an estimate of ρ_s/ρ_i, it turns out that the standard deviation of *these* is practically the same as that obtained in a simulation where ρ_s, ρ_i and A were all fixed. In other words, variation in ρ_s, ρ_i and A does not contribute to variation in the estimates of the common ratio ρ_s/ρ_i.

The idea of "matched pairs" is well known in the statistical design of experiments, and statisticians are familiar with the above phenomenon. A standard example is the "boys shoes" experiment (Box *et al.*, 1978) to compare how much two materials A and B wear. In a matched pairs design, each boy is given a pair of shoes with one shoe made of A and the other of B. This gives a much more precise estimate of the difference in wear than an experiment in which some boys have shoes made of A and other boys have shoes made of B. This is because of the large variation in wear between boys, which contributes to the comparison between A and B in the latter case, but not in the matched pairs design. Likewise in the external detector method, variation between grains does not contribute to the precision of the estimate of ρ_s/ρ_i.

3.5 Estimates of fission track age

As it stands, equation (3.6) represents an exact relation between true values. To apply it to a given sample, the quantities on the right side are either assumed known or measured experimentally and substituted to give an estimate of t. We also wish to determine the precision of this estimate. Again, it is necessary to recognise the logical distinction between the *true* fission track age t and an *estimate* of it, denoted by $\hat{t}$. In particular, we often have several different estimates of the same true age.

The quantities λ, σ_f and I are assumed to be known exactly, their values as given after equation (3.6). The neutron fluence, Φ, is usually measured indirectly by counting tracks induced in a dosimeter glass that was irradiated along with the sample. This is an artificially produced glass containing a known amount of uranium, see Section 2.7. Let N_d denote the number of these induced tracks that intersect an area A_d, where the subscript d stands for "dosimeter". Then N_d has a Poisson distribution with mean $A_d\rho_d$, where ρ_d is the true track density in the dosimeter glass. The estimated track density, $\hat{\rho}_d$, is

$$\hat{\rho}_d = \frac{N_d}{A_d} \tag{3.7}$$

and has relative standard error

$$\frac{\text{se}(\hat{\rho}_d)}{\rho_d} = \sqrt{\frac{1}{A\,\rho_d}} \approx \sqrt{\frac{1}{N_d}}. \tag{3.8}$$

In practice, N_d is nearly always greater than 2000.

There are two common approaches to using ρ_d in the fission track age equation. One is to substitute $\Phi = B_c \rho_d$ where the constant B_c is empirically determined over a number of experiments by reference to metal activation monitors of Au, Co, or Cu, for which neutron-induced beta or gamma activity is measured electronically after irradiation (Fleischer *et al.*, 1975). This gives

$$t = \frac{1}{\lambda} \log \left(1 + \frac{1}{2} \frac{\lambda}{\lambda_f} I \sigma_f B_c \rho_d \frac{\rho_s}{\rho_i} \right), \tag{3.9}$$

which is used with an assumed known value of λ_f, although not all laboratories adopt the same value.

The other approach, called the *zeta* method, is to calibrate the age equation directly by substituting $\sigma_f I \Phi / \lambda_f = \zeta \rho_d$ to give

$$t = \frac{1}{\lambda} \log \left(1 + \frac{1}{2} \lambda \zeta \rho_d \frac{\rho_s}{\rho_i} \right), \tag{3.10}$$

where ζ is a constant determined empirically by dating samples of known age (Hurford and Green, 1983). Thus for a series of age standards, equation (3.10) is used with known t to estimate ζ. This approach avoids having to adopt an explicit value for λ_f and is now the method in most common use. Formally, ζ has the units a $\times$ cm^2, or equivalently Ma $\times$ cm$^2/10^6$.

The track densities ρ_s and ρ_i are estimated from the observed numbers of spontaneous (N_s) and induced (N_i) tracks and the area (A) as

$$\hat{\rho}_s = \frac{N_s}{A} \quad \text{and} \quad \hat{\rho}_i = \frac{N_i}{A}. \tag{3.11}$$

Substituting these estimates into equation (3.10) gives the age estimate:

$$\hat{t} = \frac{1}{\lambda} \log \left(1 + \frac{1}{2} \lambda \hat{\zeta} \hat{\rho}_d \frac{N_s}{N_i} \right), \tag{3.12}$$

where, as always in this book, $\lambda = 1.55125 \times 10^{-4}$ Ma^{-1}. It is convenient to express $\hat{\rho}_d$ in units of 10^6 cm^{-2} (millions of tracks per square centimetre) so that $\hat{\zeta} \hat{\rho}_d$ is in Ma. Then $\hat{t}$ is also in Ma.

For example, for grain 1 in Table 3.1, the age estimate is

$$\hat{t} = \frac{1}{\lambda} \log \left(1 + \frac{1}{2} \times \lambda \times 350 \times 1.304 \times \frac{31}{41} \right) = 170.3 \text{ Ma.}$$

This age estimate differs from the independent age of 162 Ma. It is natural to consider whether the difference is within the expected error range.

The approximate relative standard error of $\hat{t}$ in equation (3.12) is given by

$$\frac{se(\hat{t})}{\hat{t}} \approx \left\{ \frac{1}{N_s} + \frac{1}{N_i} + \frac{1}{N_d} + \left(\frac{se(\hat{\zeta})}{\hat{\zeta}} \right)^2 \right\}^{\frac{1}{2}}. \tag{3.13}$$

This is the overall relative standard error due to estimation errors in $\hat{\rho}_s$, $\hat{\rho}_i$, $\hat{\rho}_d$ and $\hat{\zeta}$. The separate relative standard errors are added quadratically — i.e., as the square root of the sum of their squares. The first three are estimated by the approximate formulae for Poisson counts, i.e., $1/\sqrt{N_s}$, $1/\sqrt{N_i}$ and $1/\sqrt{N_d}$, as noted above. The relative standard error of $\hat{\zeta}$ is determined empirically from repeated determinations on age standards. In practice for single grains, the last two terms are usually small compared with the first two, in which case (3.13) is dominated by the sizes of N_s and N_i. Multiplying the relative standard error by $\hat{t}$ gives the approximate absolute standard error in Ma. Formula (3.13) is appropriate when the counts N_s, N_i and N_d are not too small, conventionally all greater than about 5. A method for dealing with small counts is discussed in Section 3.11.

Hence the estimated relative standard error of $\hat{t}$ for grain 1 in Table 3.1 is

$$\frac{\mathrm{se}(\hat{t})}{\hat{t}} \approx \left\{ \frac{1}{31} + \frac{1}{41} + \frac{1}{2936} + \left(\frac{10}{350}\right)^2 \right\}^{\frac{1}{2}} = 0.24\,.$$

This is a high relative standard error (24%), but typical of a single grain age estimate. It corresponds to an absolute standard error of approximately $170.3 \times 0.24 = 41$ Ma.

If we chose grain 5, the age estimate would be 219 Ma with a standard error of 34 Ma (relative standard error 15%), whereas for grain 6, the age estimate is 193 Ma with a standard error of 107 Ma (relative standard error 56%). Both the standard errors (in Ma) and relative standard errors vary substantially between grains.

On the assumption that all grains have the same true age, the common age is similarly estimated using equation (3.12), with relative standard error given by (3.13), but using the *total* numbers of spontaneous and induced tracks summed over all grains. For Table 3.1 these total numbers are 886 and 1136, respectively. Hence the common age estimate is 175.6 Ma, with a relative standard error of 5.6% and a standard error of 9.9 Ma. This estimate, obtained from the total track counts, is usually called the *pooled age*. In this example the pooled age differs from the independent age by $(175.6 - 162)/9.9 = 1.37$ standard errors. As a rule of thumb, an estimate within two standard errors of a hypothesised value is regarded as being consistent with it.

Sometimes we may want to compare estimates from two different grains with *each other* rather than with a reference value. The standard error of the difference is the square root of the sum of the squares of the separate standard errors. But it is not correct to use standard errors obtained from (3.13) for this calculation if the grains were irradiated together and counted by the same analyst. This is because the same $\hat{\zeta}$ and $\hat{\rho}_d$ apply to both grains (i.e., the same *estimates* of ζ and ρ_d) and only the components of error from $\hat{\rho}_s$ and $\hat{\rho}_i$ contribute to the comparison.

For example, the difference in the age estimates for grains 1 and 5 in Table 3.1 is $219.0 - 170.3 = 48.7$ Ma. The separate standard errors, obtained using just the first two terms of (3.13) are 40.5 Ma and 28.9 Ma, respectively. The former is close to the value of 41 Ma we found earlier, but the latter is noticeably smaller than the earlier value of 34 Ma because $1/N_s$ and $1/N_i$ are more comparable with the other terms. The standard error of the difference in the two estimates is then

$$\sqrt{40.5^2 + 28.9^2} = 49.8 \text{ Ma.}$$

The two estimates therefore differ by $48.7/49.8 \approx 1$ standard error, so they are consistent with a common true value.

3.6 Inspection of single grain data

It is useful to make a table of summary statistics, including track counts, track densities, ages, and precisions for each grain. Table 3.3 give these statistics for the data in Table 3.1.

The track densities are in millions of tracks per square centimetre, rounded to two decimal places. For example, for grain 1 the spontaneous track density is calculated as

$$\hat{\rho}_s = \frac{N_s}{A} = \frac{31}{40 \times 0.9009} = 0.86,$$

where the area of one graticule square is 0.9009×10^{-6} cm^2. Likewise, the

Table 3.3 *Single grain statistics for the apatite sample 92-1 (Table 3.1).*

Grain	N_s	N_i	Area	N_s/N_i	$\hat{\rho}_s$	$\hat{\rho}_i$	$\hat{t}$	se($\hat{t}$)	rse($\hat{t}$)
1	31	41	40	0.76	0.86	1.14	170	41	0.24
2	19	22	20	0.86	1.05	1.22	194	60	0.31
3	56	63	60	0.89	1.03	1.17	200	36	0.18
4	67	71	80	0.94	0.93	0.99	212	37	0.17
5	88	90	90	0.98	1.09	1.11	219	34	0.15
6	6	7	15	0.86	0.44	0.52	193	107	0.56
7	18	14	20	1.29	1.00	0.78	287	103	0.36
8	40	41	40	0.98	1.11	1.14	219	49	0.22
9	36	49	40	0.74	1.00	1.36	166	37	0.22
10	54	79	60	0.68	1.00	1.46	154	28	0.18
11	35	52	40	0.67	0.97	1.44	152	34	0.22
12	52	76	70	0.68	0.82	1.21	154	28	0.18
13	51	74	49	0.69	1.16	1.68	155	28	0.18
14	47	66	50	0.71	1.04	1.47	161	31	0.19
15	27	39	36	0.69	0.83	1.20	156	39	0.25
16	36	44	40	0.82	1.00	1.22	184	40	0.22
17	64	86	50	0.74	1.42	1.91	168	28	0.17
18	68	90	50	0.76	1.51	2.00	170	28	0.16
19	61	91	60	0.67	1.13	1.68	151	26	0.17
20	30	41	30	0.73	1.11	1.52	165	40	0.24
Pooled	886	1136	940	0.78	1.05	1.34	176	10	0.056

induced track density is

$$\hat{\rho}_i = \frac{N_i}{A} = \frac{41}{40 \times 0.9009} = 1.14.$$

The other track densities are calculated similarly. Also for grain 1, $\hat{t} = 170.3$ is calculated above and here rounded to 170; $\text{se}(\hat{t})/\hat{t} = \sqrt{1/31 + 1/41} = 0.24$ and $\text{se}(\hat{t}) = 170.3 \times 0.24 = 41$ to the nearest Ma. Similarly for the other grains. For these precisions we have not included the errors in $\hat{\zeta}$ and $\hat{\rho}_d$ because we usually want to compare the single grain estimates with each other. But the precision of the pooled age is calculated from (3.13), including the contributions from $\hat{\zeta}$ and $\hat{\rho}_d$.

As a general procedure this table can be inspected for unusual features, such as extreme ratios, high or low track densities, low or high precision grains and outlying ages. For this purpose it is important not to tabulate too many decimal places — two are usually appropriate for individual track densities and relative standard errors. In Table 3.3 the data appear to be very clean. The relative standard errors, rse$(\hat{t})$, for most of the single grain ages are around 20% and the relative standard error for the pooled age is 5.6%.

There is another feature that is worth pointing out, illustrated by grains 4 and 19. These have the same relative standard error (17%), but grain 4, with a larger estimated age, has a larger standard error in Ma. This happens generally and should be appreciated when interpreting precisions of age estimates. The track counts N_s and N_i are comparable in size for these two grains, producing the same relative standard error. By contrast, grain 15 has a similar absolute standard error (39 Ma) to that of grain 4 (37 Ma), but its relative standard error is higher (25%) because N_s and N_i are smaller. Likewise, the track counts for grain 7 are more than double those for grain 6, so its relative standard error is much smaller. But these grains have quite similar absolute standard errors because the ratio N_s/N_i, and hence the age estimate, for grain 7 happens to be higher. In general, the sizes of the counts determine the *relative* standard error. The absolute standard error also depends on the size of the estimate, which is related to the ratio of the counts.

3.7 Radial plot of single grain ages

To compare a number of single grain fission track age estimates it can be useful to plot them in a graph. A histogram has sometimes been used for this purpose, but this is a poor graphical method for fission track ages. It is hard to interpret, because each estimate has a different precision and the precisions usually vary substantially. For example, if you see a large extreme value in the histogram, you cannot tell whether it may represent a larger true value or merely a moderate true value but with a large standard error. Moreover, as noted above, larger fission track age estimates tend to have larger standard errors, even if their relative standard errors are the same. This has the effect of making a histogram look *positively skewed* — larger values being more spread than smaller ones. So the shape of the histogram is dictated very much by the estimation *errors* and not necessarily by the fission track ages themselves.

The latter feature can be improved by plotting the estimates on a logarithmic scale, or on some other transformed scale where the standard errors do not depend on the sizes of the estimates. But the problem that the estimates have different precisions still remains. A variant of a histogram, sometimes called a probability density plot or weighted histogram, has similar drawbacks, see the Appendix, Section A.9.

In fact, because their precisions differ, it is inherently not straightforward to compare single grain age estimates. A *radial plot* is a valid method for doing so. The principle behind it is described in the Appendix, Section A.8, and some applications to fission track ages are given in Galbraith (1990). For several estimates $z_1, z_2, \ldots, z_n$ with standard errors $\sigma_1, \sigma_2, \ldots, \sigma_n$, a radial plot is a scatterplot of the standardised estimate y_j against the precision x_j, where

$$y_j = \frac{z_j - z_0}{\sigma_j}, \qquad x_j = \frac{1}{\sigma_j} \tag{3.14}$$

and z_0 is a suitable reference value.

There are several alternative methods of constructing a valid radial plot. For example we could take z_j to be the fission track age estimate in Ma, or its natural log, or indeed some other transformed estimate. A reason for using a transformed scale is produce a graph that is not dominated by the relation between the size of the estimate and its standard error and is therefore easier to interpret.

The following method, using the *angular* transformation for binomial counts, deals with this problem and also copes well with small numbers of spontaneous tracks, including zero, as well as large counts. Let

$$z_j = \arctan \sqrt{\frac{N_{sj} + \frac{3}{8}}{N_{ij} + \frac{3}{8}}} \tag{3.15}$$

for $j = 1, 2, \ldots, n$, where N_{sj} and N_{ij} denote the numbers of spontaneous and induced tracks for grain j. This transformation is often expressed in the equivalent form

$$z_j = \arcsin \sqrt{\frac{N_{sj} + \frac{3}{8}}{N_{sj} + N_{ij} + \frac{3}{4}}} \, .$$

Adding $\frac{3}{8}$ to each count in (3.15) is a correction proposed by Anscombe (1948) that is useful for small and zero counts and that has negligible effect for large counts. The approximate standard error of z_j is then

$$\sigma_j = \frac{1}{2} \left(\frac{1}{N_{sj} + N_{ij} + \frac{1}{2}} \right)^{\frac{1}{2}} \qquad j = 1, 2, \ldots, n, \tag{3.16}$$

which takes into account estimation errors from $\hat{\rho}_s$ and $\hat{\rho}_i$ only. As mentioned earlier, the errors in $\hat{\zeta}$ and $\hat{\rho}_d$ are common to all grains and do not contribute to comparisons between them.

A suitable value of z_0 in (3.14) may be chosen according to circumstances. A convenient default value is

$$z_0 = \arctan \sqrt{\frac{N_s}{N_i}}, \tag{3.17}$$

where, as before, N_s and N_i denote the total spontaneous and induced track counts summed over all grains in the sample. This corresponds to the pooled age transformed to the angular scale.

Figure 3.2 shows a radial plot of the data in Table 3.1 using this method with $z_0 = 0.72342$ radians, obtained from equation (3.17). A single grain age can be read off by extrapolating a line from $(0,0)$ through (x_j, y_j) to the age scale. This is illustrated for grain 5, plotted as an open circle; the reader may verify from (3.15) that $z_j = 0.77980$ and from (3.16) that $\sigma_j = 0.03742$, and hence $x_j = 26.72$ and $y_j = 1.507$. The slope of the line through this point is $y_j/x_j = z_j - z_0 = 0.0564$. This corresponds to an age of

$$\hat{t}_j \;=\; \frac{1}{\lambda} \log \left(1 + \frac{1}{2} \lambda \hat{\zeta} \hat{\rho}_d \tan^2 z_j \right) \;=\; 219 \text{ Ma}.$$

Grain 8, plotted as an open triangle, has an almost identical z_j, and hence has the same estimated age. Both grain 5 and grain 8 lie on the line joining the origin to age 219 Ma. Points lying on the same radius have the same estimated fission track age.

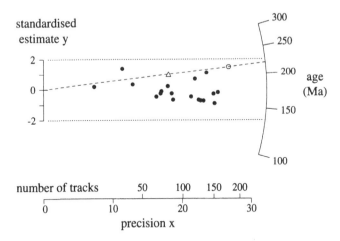

Figure 3.2 *Radial plot of fission track age estimates for the apatite sample 92-1 (Table 3.1). This is a scatterplot of the points (x_j, y_j) in equations (3.14) and (3.17). Each estimate has unit standard error on the y scale. The x-axis also shows the total numbers of tracks $N_{sj} + N_{ij}$ per grain. Ages are read off by extrapolating a line from $(0,0)$ through a plotted point to the age scale. Because the points all lie between -2 and $+2$ on the y scale, they are consistent with the pooled age of 175.6 Ma.*

The age scale is calculated by choosing values of t (here 100, 150, 200, 250 and 300) and using the equation

$$z = \arctan \sqrt{\frac{e^{\lambda t} - 1}{\frac{1}{2}\lambda \hat{\zeta} \hat{\rho}_d}} \tag{3.18}$$

to obtain the corresponding values of z, and hence to obtain corresponding slopes $z - z_0$ of the lines through $(0,0)$.

When the data are consistent with a common age, and the default z_0 is used, the points should scatter homoscedastically about the horizontal line through $(0, 0)$. Furthermore, because the y_j have unit standard deviation, about 95% of the points should lie in a band between -2 and $+2$ on the y scale. This pattern is evident in Figure 3.2, where *all* points lie well within this band. Estimates that have high precision lie towards the right end of the graph, and those with low precision lie nearer the origin.

Sometimes it is convenient to center the plot at a value other than the pooled age (i.e., to use a different z_0) — for example, at an independently determined age, or at the pooled age of several samples when comparing them on the same scales. Figure 3.3 shows the data from Table 3.1 centered at 100 Ma, for illustration. The points now lie within a ± 2 band around the radial line corresponding to the pooled age. The equation for calculating z_0 in terms of a reference age of t_0 Ma is

$$z_0 = \arctan \sqrt{\frac{e^{\lambda t_0} - 1}{\frac{1}{2}\lambda \hat{\zeta} \hat{\rho}_d}}. \tag{3.19}$$

Hence for $t_0 = 100$ Ma, $\hat{\zeta} = 350$, $\hat{\rho}_d = 1.304$, we find that $z_0 = 0.5865$.

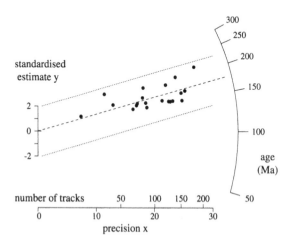

Figure 3.3 *Another radial plot of the data in Table 3.1, but with the age scale centered at 100 Ma. The points scatter within a ± 2 band about the radial line pointing to the pooled age of 175.6 Ma.*

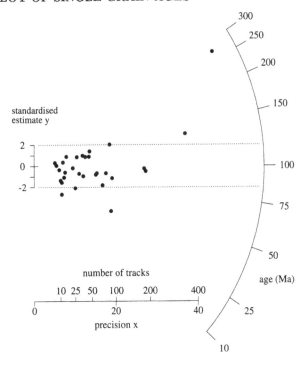

Figure 3.4 *Radial plot of fission track ages from the data in Table 3.2 (Mahe granite, Seychelles). The points do not lie in or near a ±2 band about a radial line and the ages are therefore not consistent with a common value.*

Figure 3.4 shows a radial plot of the data in Table 3.2. It is clear that the single grain age estimates vary widely and are not consistent with a common value. Six of the 30 points lie further than two standard errors from the pooled age, outside the horizontal lines. There is one very old grain with high precision, but even if this is excluded, the remaining single grain ages are over-dispersed.

Data may depart from homogeneity in several ways. There may be outliers (points that do not conform to the general pattern of the data), grains from more than one source, experimental artefacts, or varying chemical compositions between grains that have experienced thermal annealing. Such departures can often be diagnosed from a radial plot.

With respect to Figure 3.2, it is worth remarking that, because the standard errors are based solely on the Poisson distributions of the counts, the fact that all of the estimates agree within error lends further support to the Poisson line segment model. The fission track counts do vary consistently with Poisson variation, once the variation in uranium is eliminated. Consequently also, over-dispersion, such as in Figure 3.4, may usually be regarded as genuine and not just "experimental error".

3.8 Chi-square age homogeneity test

A radial plot provides a visual assessment of the spread of individual grain ages. In Figure 3.2 it is clear that the grain ages are homogeneous, whereas in Figure 3.4 they are clearly heterogeneous. For some data sets, it is not immediately obvious whether the single grain estimates are consistent with a common true age. In such cases a χ^2 test may help to answer this question. This is a standard test for homogeneity in a $n \times 2$ table of counts (e.g., Snedecor and Cochran, 1980, Section 11.7; Armitage et $al.$, 2002, Section 8.5), which is also applicable to matched pairs of Poisson counts. It differs from the test used in Section 2.7. There, we were assessing whether the variation between several counts was consistent with Poisson variation, or equivalently, whether the counts were from Poisson distributions with a common mean. Here the pairs of counts are from different Poisson distributions, but we are assessing whether the ratio of true means ρ_s/ρ_i is the same for each pair.

To apply this test, calculate the χ^2-statistic given by

$$\chi^2_{\text{stat}} = \frac{1}{N_s N_i} \sum_{j=1}^{n} \frac{(N_{sj} N_i - N_{ij} N_s)^2}{N_{sj} + N_{ij}}. \tag{3.20}$$

If ρ_s/ρ_i is the same for each grain, then this will be a value from a χ^2 distribution with $n-1$ degrees of freedom. The larger the value of χ^2_{stat}, the more evidence there is against this hypothesis. Significance is assessed by calculating the p-value, which is the probability that a value from the $\chi^2(n-1)$ distribution exceeds χ^2_{stat}. A small p-value is evidence that the data are not consistent with a common ratio ρ_s/ρ_i. As a general convention, a p-value below 0.05 is regarded as moderate evidence against the null hypothesis, and a p-value below 0.01 is strong evidence. Some notes on significance tests and p-values are given in the Appendix, Section A.7.

For the data in Table 3.1, $\chi^2_{\text{stat}} = 10.94$ with 19 degrees of freedom. The p-value is therefore the probability that a value from a $\chi^2(19)$ distribution exceeds 10.94. This may be found from statistical tables or software and equals 0.93, which implies that the data are consistent with a common ρ_s/ρ_i, and hence with a common fission track age. For Table 3.2, $\chi^2_{\text{stat}} = 152.0$ with 29 degrees of freedom, and the p-value is considerably less than 0.001, which is strong evidence that the true fission track ages are not all equal. In each case, the χ^2 test confirms the visual impression from Figures 3.2 and 3.4.

When the track counts are small, the χ^2-statistic does not have a χ^2 distribution, even approximately. A common rule of thumb for using the p-value obtained from the χ^2 distribution is that the $expected$ frequencies, given by $N_s(N_{sj} + N_{ij})/(N_s + N_i)$, should be greater than 5 for all grains, although some leeway is allowable. When there are several small expected frequencies, one can still calculate the χ^2-statistic, but a more accurate method of assessing its significance may be needed. The "exact" p-value can be computed from the appropriate multiple hypergeometric distribution. Agresti (1996, p. 44) gives a simple example. There is statistical software available to carry out this calculation, for example, StatXact (CYTEL Software Corporation, 1991) or S-plus (Statistical Sciences, 1992).

The χ^2-statistic is a simple diagnostic. If it is significantly large (i.e., if the p-value is small) this is evidence that the true ratios ρ_s/ρ_i for the individual grains, and hence their true fission track ages, differ. If it is not significantly large, the data are reasonably consistent with a common ratio ρ_s/ρ_i. This is of course not the same as saying that they really do have the same ρ_s/ρ_i. It sometimes happens that the χ^2-statistic is not significantly large simply because there is not enough information in the data to detect differences in ρ_s/ρ_i — for example, when the track counts are small or when the amount of variation of the true ages is small. It is therefore useful also to *estimate* how dispersed the true fission track ages might be.

3.9 A measure of age dispersion

In many situations the true ratios ρ_s/ρ_i vary between grains. For example, tracks may have been shortened by heat but by differing amounts in different grains because of their differing chemical compositions. So we may have a *distribution* of true fission track ages, rather than a single common value, to estimate.

The contrasting situations are illustrated in Figure 3.5. The assumption that all grains have the same true age is depicted by a solid vertical line at the common value of 126 Ma. If we take a sample of 10 grains from this population, each grain will have the common age. This is illustrated by a stack of 10 open circles at 126 Ma. When these grains are dated, their estimated ages will vary, because of Poisson variation in track counts, as seen in the dot plot of estimated ages and in the radial plot, here drawn vertically. The radial plot confirms that the estimated ages look to be consistent with a common value.

An alternative assumption is to suppose that the true fission track ages vary according to some statistical model. In the random effects model, $\log(\rho_s/\rho_i)$ is drawn from a normal distribution with mean μ and standard deviation σ. This normal curve is drawn in Figure 3.5, and the corresponding non-linear age scale is marked. Now when a sample of 10 grains is taken, their true fission track ages vary, as illustrated by the dot plot of open circles. When these grains are dated, the estimates vary more widely, because the Poisson variation in track counts is superimposed onto the variation of true values, as seen in the dot plot of estimated ages. The radial plot captures the over-dispersion in estimates.

To say that $\log(\rho_s/\rho_i)$ has a normal distribution means that ρ_s/ρ_i has a *log*-normal distribution, which has a positively skewed shape. The distribution looks symmetrical in Figure 3.5 because it is drawn on a log scale. The age corresponding to μ is known as the *central age*. It corresponds to the population mean of $\log(\rho_s/\rho_i)$ and hence to the geometric mean of ρ_s/ρ_i, and also to the population median of either. The parameter σ is known as the *age*

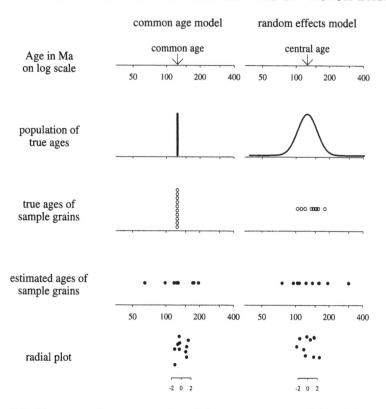

Figure 3.5 *Diagrammatic representation of the common age model and the random effects model showing typical samples of true fission track ages, estimated ages and radial plots.*

dispersion. It is the population standard deviation of values of $\log(\rho_s/\rho_i)$. It is also approximately the *relative* standard deviation of the true fission track ages. When $\sigma = 0$ the model reduces to the common age model.

In Figure 3.5, μ is such that the central age is 126 Ma and the value of σ is 0.2, corresponding to a 20% relative standard deviation of true fission track ages. Thus the middle 95% of true fission track ages in the population of grains would be in the range $126 \times \exp(\pm 2 \times 0.2)$, i.e., from 84 Ma to 188 Ma.

In practice one will have observed counts from a sample of grains from which one may wish to estimate μ and σ. The estimates will differ from the true values for two reasons: (a) one only has a sample from the population of grains and (b) for any grain in the sample one only has an estimate of ρ_s/ρ_i, rather than the true value. The precisions of $\hat{\mu}$ and $\hat{\sigma}$ will reflect both sources of variation. It is not possible to give reliable explicit formulae to estimate μ and σ, but we give here a simple numerical algorithm that is easily programmed. The random effects model is discussed in more detail in Chapter 6, Section 6.3.

3.9.1 A simple algorithm to estimate the age dispersion and central age

Let $N_{sj}, N_{ij}, j = 1, 2, \ldots, n$ be spontaneous and induced track counts for n grains, generated according to the random effects model. Also let $m_j = N_{sj} + N_{ij}$ and $p_j = N_{sj}/m_j$.

1. Set convenient starting values to compute σ and θ iteratively. For example, $\sigma = 0.15$ and
$$\theta = \frac{\sum_{j=1}^{n} N_{sj}}{\sum_{j=1}^{n} m_j}.$$

2. For $j = 1, 2, \ldots, n$ compute
$$w_j = \frac{m_j}{\{\theta(1 - \theta) + (m_j - 1)\theta^2(1 - \theta)^2\sigma^2\}}.$$

3. Compute new values of σ and θ as
$$\sigma = \sigma \left(\frac{\sum_{j=1}^{n} w_j^2 (p_j - \theta)^2}{\sum_{j=1}^{n} w_j} \right)^{\frac{1}{2}} \quad \text{and} \quad \theta = \frac{\sum_{j=1}^{n} w_j p_j}{\sum_{j=1}^{n} w_j}.$$

4. Repeat steps 2 and 3 until σ and θ do not change; this is usually achieved within 30 iterations.

5. The final value of σ, denoted by $\hat{\sigma}$, is the estimated age dispersion, and the final value of θ, denoted by $\hat{\theta}$, is the estimate of $\rho_s/(\rho_s + \rho_i)$. The central age t_μ is then estimated from the age equation as
$$\hat{t}_\mu = \frac{1}{\lambda} \log \left(1 + \frac{1}{2}\lambda\hat{\zeta}\hat{\rho}_d \frac{\hat{\theta}}{1 - \hat{\theta}} \right)$$

and its relative standard error is estimated as
$$\frac{se(\hat{t}_\mu)}{\hat{t}_\mu} = \left[\frac{1}{\hat{\theta}^2 (1 - \hat{\theta})^2 \sum_{j=1}^{n} w_j} + \frac{1}{N_d} + \left(\frac{se(\hat{\zeta})}{\hat{\zeta}} \right)^2 \right]^{\frac{1}{2}}$$

which includes contributions due to error in estimating ρ_d and ζ.

This algorithm gives a precision for the central age estimate but not for the age dispersion. When $\sigma = 0$, $\hat{t}_\mu$ is the pooled age, and its relative standard error reduces to the usual formula (3.13). The random effects model may also be fitted by maximum likelihood (see Chapter 6).

Consider the data in Table 3.1, which from the radial plot and χ^2 homogeneity test, are consistent with the common age model. Here, this algorithm converges to $\hat{\sigma} = 0$ and $\hat{t}_\mu = 176$ Ma with standard error 9.9 Ma, which agrees with the pooled age calculated in Section 3.5. So here the algorithm has effectively fitted the common age model.

For the data in Table 3.2, which are not consistent with the common age model, the algorithm gives $\hat{\sigma} = 0.446$, indicating an estimated 45% relative standard deviation in the ages of the underlying population of grains. The estimated central age is $\hat{t}_\mu = 94.1$ Ma with standard error 10.1 Ma. The corresponding maximum likelihood estimates (see Chapter 6, Section 6.4) are $\hat{\sigma} = 0.443$ (with standard error 0.090) and $\hat{t}_\mu = 90.5$ Ma (with standard error 10.6 Ma), which are practically the same.

There are a variety of geological processes that lead to over-dispersion in fission track age estimates. Consequently, there are a variety of statistical models that might be used. Often the geological context is not sufficiently well understood to determine what model to fit. So the analyst may explore several alternative models, such as finite mixtures, random effects, or generalisations of these that are considered in Chapters 5 and 6.

3.10 A protocol for data analysis

Proper statistical data analysis is usually not automatic, but rather involves interaction between the data and analyst in a given context. Nevertheless it is useful to have a routine protocol at least for initial data analysis. Such a protocol might include the following steps, as indicated in the preceding sections:

1. Inspection of single grain data.
2. Radial plot of single grain ages.
3. Chi-square homogeneity test.
4. Calculation of pooled age and its precision.
5. Calculation of age dispersion and central age.

Further analysis and presentation would depend on what is learnt after completion of these steps.

3.11 Dealing with small counts

We noted earlier that formula (3.13) is valid provided the expected track counts are greater than about 5. However, spontaneous track counts are often smaller, even 0 or 1. For example, if $N_s = 0$, equations (3.12) and (3.13) formally give an age estimate of zero with infinite relative standard error, which is neither useful nor appropriate. Even if $N_s = 1$ or 2, the age estimate and relative standard error are unreliable. In such cases, precision is better indicated by calculating a confidence interval. To do this, calculate upper and lower confidence limits for ρ_s/ρ_i and insert these in place of N_s/N_i in equation (3.12).

Such confidence limits may be calculated by adapting standard confidence intervals for a binomial parameter θ (e.g., Armitage et al., 2002, p. 117) and using the equation

$$\frac{\rho_s}{\rho_i} = \frac{\theta}{1 - \theta} \tag{3.21}$$

as described in the Appendix, Section A.3.

The upper confidence limit for θ, denoted by θ_U, is given by

$$\theta_U = \frac{x_\alpha(N_s + 1)}{x_\alpha(N_s + 1) + N_i}, \tag{3.22}$$

where x_α is the upper α percentage point of the F-distribution with degrees of freedom $2N_s + 2$ and $2N_i$.

Similarly, the lower confidence limit for θ, denoted by θ_L, is

$$\theta_L = \frac{N_s}{N_s + y_\alpha(N_i + 1)}, \tag{3.23}$$

where y_α is the upper α percentage point of the F-distribution with degrees of freedom $2N_i + 2$ and $2N_s$. Tables of percentage points of the F-distribution are widely available and are programmed into most statistical software packages.

To illustrate the calculation of an upper 95% confidence limit when $N_s = 0$, consider grain 1 in Table 3.2, where $N_s = 0$ and $N_i = 11$. The degrees of freedom are $(2 \times 0) + 2 = 2$ and $2 \times 11 = 22$. From F-tables, the upper 5% point of an $F(2,22)$ distribution is 3.443. Hence

$$\theta_U = \frac{3.443 \times (0 + 1)}{3.443 \times (0 + 1) + 11} = 0.2384.$$

So the upper confidence limit for ρ_s/ρ_i from (3.21) is $0.2384/(1 - 0.2384) = 0.313$. Hence the upper confidence limit for the age is

$$t_U = \frac{1}{\lambda} \log\left(1 + \frac{1}{2} \times \lambda \times 380 \times 1.257 \times 0.313\right) = 74 \text{ Ma}.$$

When $N_s = 0$, the lower confidence limit is taken to be 0 Ma. The interval from 0 to 74 Ma is therefore a 95% confidence interval for t.

To illustrate the calculation of a 95% confidence interval when N_s is small but not zero, consider grain 11, where $N_s = 1$ and $N_i = 9$. We calculate upper and lower 2.5% limits using equations (3.22) and (3.23).

For the upper limit, the degrees of freedom are $(2 \times 1) + 2 = 4$, and $2 \times 9 = 18$. From F-tables, the upper 2.5% point is 3.608. Hence

$$\theta_U = \frac{3.608 \times (1 + 1)}{3.608 \times (1 + 1) + 9} = 0.4450.$$

The upper confidence limit for ρ_s/ρ_i from (3.21) is $0.4450/(1 - 0.4450) = 0.802$. Hence the upper confidence limit for the age is

$$t_U = \frac{1}{\lambda} \log\left(1 + \frac{1}{2} \times \lambda \times 380 \times 1.257 \times 0.802\right) = 189 \text{ Ma}.$$

For the lower limit given by (3.23), the degrees of freedom are $(2 \times 9) + 2 = 20$ and $2 \times 1 = 2$. From F-tables, the upper 2.5% point is 39.45. Hence

$$\theta_L = \frac{1}{1 + 39.45 \times (9 + 1)} = 0.002528.$$

The lower confidence limit for ρ_s/ρ_i from (3.21) is $0.002528/(1 - 0.002528) =$

0.00253. Hence the lower confidence limit for the age is

$$t_L = \frac{1}{\lambda} \log \left(1 + \frac{1}{2} \times \lambda \times 380 \times 1.257 \times 0.00253 \right) = 0.6 \text{ Ma.}$$

A 95% confidence interval for t therefore goes from 0.6 to 189 Ma — a rather wide interval, but not necessarily completely uninformative. The statistical software StatXact (CYTEL Software Corporation, 1991) includes calculation of θ_U for $N_s = 0$ and (θ_L, θ_U) for $N_s > 0$, as above.

The case $N_s = 0$ deserves special comment as it is quite common in practice and often misunderstood. When $N_s = 0$, equation (3.12) formally estimates t to be zero, with infinite relative standard error given by equation (3.13), which is not sensible. To illustrate the problem, suppose we saw a grain with $N_s = 0$ and $N_i = 5$. This might be a young grain or it might simply have a low uranium content and possibly be quite old. On the other hand if we saw $N_s = 0$ with $N_i = 50$ we would be more inclined to believe that the grain was young or had been heavily annealed, because there is clearly a reasonable amount of uranium present. The same applies for small non-zero numbers of spontaneous tracks, albeit to a lesser extent. This difference in information is not properly reflected in formulas (3.12) and (3.13), which are applicable for moderately large counts. The confidence limits (3.22) and (3.23) do reflect the information correctly and become smaller as N_i increases. The same applies to σ_j given by (3.16).

The χ^2 homogeneity test is also affected by small counts, as discussed in Section 3.8. In this situation it may help to estimate a confidence interval for the age dispersion. A more extensive example with small spontaneous track counts is discussed in Chapter 5, Section 5.4.

3.12 Practical considerations

It is important for anybody interpreting fission track data to recognise that formal statistical analysis is based on an idealisation of the way in which tracks are generated and recognised by the observer. It is also important to understand, if only qualitatively, the way in which departures from the ideal might affect the analysis in practice. Some of these potential problems are encountered only in the external detector method, while others apply to all fission track dating attempts.

3.12.1 Sample collection, rock crushing and separation of grains

Ideal: Uncontaminated samples are collected. Samples are crushed, liberating whole grains, which are then put through a laboratory flotation process, creating pure separates of apatite, zircon or other minerals of interest by bracketing the mineral's density with heavy liquids.

In practice: Poor field collection may mean that the sample contains material from other sources. Sometimes the introduction of such material is inevitable, for example in a drilling mud.

Some grains may be broken up by crushing and appear as several grains in the final sample.

Variations in density of grains, particularly due to imperfect liberation or composite grains, may mean that the bracketing densities have to be fairly broad, causing the sample to include other minerals than the one of interest.

Poor laboratory practice, such as the use of Teflon stoppers in separating funnels, may cause contamination by the carry-over of trapped grains from one sample to the next.

3.12.2 Grain selection and counting of tracks

Ideal: The observer selects a random sample of say 50 or 100 grains for track counting. These are perfectly polished and etched, with high etching efficiency. In apatite, grains with a polished surface parallel to the crystallographic c-axis (prismatic grains) have high etching efficiency and are preferred over basal grains. Prismatic grains have track openings aligned in the direction of the c-axis.

All of the randomly chosen grains are used. All spontaneous tracks are recognised and counted, and no defects are included. Induced tracks are counted in exactly the same matched area as for spontaneous tracks, so that induced track counts in the detector accurately reflect the uranium content of this area of the grain.

Uranium atoms are homogeneously distributed across both sides of the polished grain surface.

In practice: Some rocks are low yielding, and the observer may have to make do with as few as 4 or 5 grains. In some laboratories, the law of diminishing returns means that a maximum sample size of 20 grains is used in routine work.

In low-yielding samples, grain selection may have to be expanded to include slightly non-prismatic grains.

Some grains have an extremely high density of spontaneous tracks, which is impossible to count. The most the observer can do is to record the presence of such grains.

It is possible even for experienced counters to have difficulty distinguishing between dislocations and tracks. In some samples, particularly rapidly cooled ones, virtually all grains have dislocations, and it is not possible to exclude such grains. In samples from mixed sources, sampling only grains without dislocations may cause bias.

Many grains contain fractures, and highly fractured grains are usually excluded, but there is a question of whether to count across small fractures.

When counting in apatite, some grains may contain zircon inclusions or be mounted adjacent to one or more zircons, which may produce spurious induced tracks in the mica.

Poor contact between grain and mica may cause the mica image to be larger than the polished grain outline, which can make it difficult to locate the image of the edge of the grain in the detector and can cause counting bias at the edges. Most laboratories have a well-defined protocol for deciding whether a grain is suitable. This may involve selecting an area well inside the grain, or even rejecting that grain if the problem is too severe.

When the uranium concentration is low, it can be difficult to match the grain exactly with its mica image. Also, if the uranium is distributed heterogeneously in the direction perpendicular to the polished plane, the track densities ρ_s and ρ_i may not be correctly matched.

These problems mean that there is inevitably some subjectivity in the selection of grains and counting of tracks. Therefore careful training and monitoring of different observers and laboratories is an important part of practical fission track analysis.

3.13 Remarks

This chapter has derived the mathematical basis of fission track dating and discussed the initial statistical analysis of the data, concentrating on the most popular experimental design, the external detector method. Formulae are given for age estimates and their precisions when the grains from a single sample have a common age, and for a measure of age dispersion when there is a spread in ages. Other statistical methods and measures are possible, but, for definiteness, we have presented methods that have proved useful over a number of years and are generally robust. The next chapter sets out corresponding statistical models and analyses for another experimental design, called the *population method*.

Both of these chapters deal with the dating of grains from a single sample. This is a basic step in any routine analysis, although most geological applications require the dating of a suite of related samples. Such samples may have one or more parameters in common, or the ages may display a meaningful pattern with co-variables of geological interest, such as depth down a borehole or spatial position across a geological regime. Chapters 5 and 6 address some of the statistical problems posed by the modelling of such data.

When ages of different samples are to be compared, some sources of variation can be avoided by sensible experimental design. For example, if different samples are irradiated in the same canister, they will receive the same dose (or nearly the same dose if there is a gradient in the thermal neutron fluence) and hence the error associated with $\hat{\rho}_d$ does not affect the comparison. Also the error in $\hat{\zeta}$ is avoided when comparing fission track dates from different samples counted by the same analyst.

The precisions of age estimates presented here take into account only natural sources of variation of track counts and the zeta calibration. However, a series of inter-laboratory comparisons undertaken in conjunction with the International Fission Track Workshops held once every 4 years (for example, Miller et al., 1993) suggest that further sources of variation may exist. Statistical analysis of these trials reveals that between-laboratory variation is sometimes of the same order of magnitude as the natural variation. This is relevant to geologists attempting to interpret fission track ages from samples dated by different laboratories.

Finally, and perhaps most importantly, modern fission track analysis of apatites involves inferences from both counts and lengths. In some situations, ignoring lengths can be misleading. Chapters 7 and 8 deal with theoretical and observational features of fission track lengths.

3.14 Historical note

Many of the statistical formulae in this chapter have been used routinely in laboratories for some years. In the early 1980s a controversy arose about the use of standard errors based on the Poisson distribution, as in equation (3.13). A leading laboratory had noticed that, when repeatedly dating sets of grains from the same large sample (which was an age standard, with a known age) the age estimates appeared to be more accurate than this formula predicted. That is, the age estimates obtained did not vary enough! An alternative formula was put forward (McGee and Johnson, 1979) that produced smaller values for the standard errors and that appeared to agree with the variation in the observed estimates. This was based on a well-known approximate formula for the variance of a ratio Y/X of random variables (e.g., Rice, 1995, page 153), viz.,

$$\mathrm{var}\left(\frac{Y}{X}\right) \approx \frac{\mu_Y^2}{\mu_X^2}\left\{\frac{\sigma_Y^2}{\mu_Y^2} - 2\rho_{XY}\frac{\sigma_Y}{\mu_Y}\frac{\sigma_X}{\mu_X} + \frac{\sigma_X^2}{\mu_X^2}\right\},$$

where μ_X, σ_X, μ_Y and σ_Y are the means and standard deviations of X and Y and ρ_{XY} is their correlation coefficient. This formula was applied with $Y = N_s$ and $X = N_i$, so that the square root of the term in braces was effectively the relative standard error of the estimate of ρ_s/ρ_i. Values of $1/\sqrt{N_s}$ and $1/\sqrt{N_i}$ were substituted for the relative standard deviations σ_Y/μ_Y and σ_X/μ_X and a value for the correlation ρ_{XY} was found by calculating the empirical correlation from pairs of counts (N_s, N_i) for different grains. But this correlation, which is noted in Section 3.4, arises when comparisons are made between grains, while these relative standard deviations are based on the Poisson distribution, and only reflect variation "within" grains. The new formula was therefore incorrect.

Counter arguments were put by Green (1981a) and Galbraith (1981), but many practitioners were unable to see which were right, especially as the new formula appeared to explain the above phenomenon. The situation was made more confusing because when the correlation ρ_{XY} was set to zero then the formula gave the usual (right) answer — and the numbers of spontaneous and induced tracks are obviously correlated! The argument called into question the basis of the Poisson model, the isotopic ratio and other aspects that were important but not directly relevant to the specific statistical question. McGee *et al.* (1985) later found, by doing simulations, that if the "between grain" variances of N_s and N_i were used (and if there were a reasonably large number of grains so that they were well-estimated, and if the areas counted did not vary much) then the new formula gave results that agreed closely with (3.13). But even in these favourable cases, this formula is inefficient in the present context and is not recommended.

The phenomenon of correlation in paired data has also caused confusion in medicine and in other fields, and there is more than one way to view it. There are wider difficulties too that are highlighted by the above and similar episodes. One concerns the use of mathematics. For example, how can one doubt such a well-known formula as that above, which is quite general and requires no distributional assumptions? But the real issues are usually the applicability of the mathematics to the situation at hand and the "real world" meanings of the terms used. These are often not explicit and it can be hard to judge what is appropriate. Another concerns language and terminology. There were a number of technical statistical concepts that were relevant to the discussion, including sampling distributions, conditional probability, blocking (or matching) in experimental designs, within and between-unit variation, and the important distinction between a "true" quantity and an estimate of it. These are familiar to statisticians, but they are not part of the usual language and notation in applied science. The use of simulations can sometimes help to clarify ideas, although care is needed that fallacies are not built into the simulation program.

Interestingly, the laboratory's estimates really were under-dispersed, and we were left without an explanation (Galbraith, 1986), though a clue to this was eventually suggested by splitting each sample into two halves. It was found that within each half the estimates varied by about the right amount, but that estimates from the first half were negatively correlated with those from the second half (Galbraith, 1990). This is strange because the errors from separate dating attempts should be uncorrelated, and those within the same sample should be slightly positively correlated if the grains were irradiated together. Such a phenomenon could happen inadvertently when repeatedly analysing the same sample, especially if analyses are not done "blind". For the analyst may know what answer to expect and sub-consciously favour that, for example by rejecting or re-measuring observations that look discrepant. There are some well-known cases of scientific data being "too good", perhaps the most famous being Mendel's genetic data on sweet peas (Fisher, 1936).

CHAPTER 4

The population method

This is an alternative experimental design to the external detector method. In statistical terms, counts of spontaneous and induced tracks are obtained from two independent samples of grains, rather than as matched pairs. Use of the population method is limited almost exclusively to apatite for reasons discussed below. A variant of it, the population-subtraction method, is used for dating volcanic glasses. Here we present some simple statistical analyses for routine use, analogous to those in Chapter 3. As always, the data need to be considered in their geological setting, and further analyses may be warranted.

4.1 Experimental method and data

After separation from the host rock, the sample of mineral grains is split into two groups. Grains from one group are polished and etched in the same way as for the external detector method to reveal spontaneous tracks. Grains from the other group are heated to remove all spontaneous tracks and are then irradiated, polished and etched to reveal induced tracks created during irradiation. Thus the induced fission tracks are etched and counted in the apatite crystals, rather than in a mica detector as they would be for the external detector method.

The two groups are often referred to as the "spontaneous" and "induced" groups, respectively. It is important for the validity of the method that the split of the sample is done *randomly*, so that each grain has the same chance of being allocated to the spontaneous group. This is especially true if the uranium concentrations differ in different grains, which they usually do. A random split where each grain has an equal chance of being allocated to either group would produce approximately equal-sized groups, which is a good thing if the average number of tracks counted per grain is similar for each group.

When counting tracks, the same area A of crystal surface is normally used for every grain, whichever group it is in. Although this is not strictly necessary, it is to be preferred in order to have confidence in the assumptions underlying the age calculation. If A was the same for grains within each group, but different for the two groups, then some further calibration might be necessary. If A varied between grains generally, then a further assumption about how areas are related to track densities is really needed to justify the calculation.

When uranium concentrations vary between grains, there is a trade-off between the numbers of tracks counted per grain and the number of grains. It is important that the latter is large enough in both groups. Sometimes a high

57

Table 4.1 *Spontaneous and induced track counts determined by the population method for the apatite sample 92-1 (Miller et al., 1993).*

Number of fission tracks counted per grain

Spontaneous	40	56	58	44	45	36	37	23	14	29	44
	36	45	49	45	45	61	33	38	50	51	34
	24	36	39	25	31	42	26	31	31	32	39
	19	34	36	23	35	30	36	26	29	38	46
	31	27	26	41	30	30	21	23			
Induced	10	13	16	11	10	18	16	12	11	12	7
	8	10	14	12	12	7	12	19	16	9	20
	10	13	15	8	7	10	11	13	8	13	12
	14	14	9	8	11	8	11	13	5	9	14
	5	15	13	12							

Counting area per grain $A = 65 \times 10^{-6}$ cm^2; $\hat{\zeta} = 351$, se$(\hat{\zeta}) = 6$ Ma/10^6; $N_d = 3000$ tracks; $\hat{\rho}_d = 0.14 \times 10^6$ tracks/cm^2. See the text for the origin of these values.

neutron fluence in the irradiation may produce much higher track densities in the induced group. Then it would be more efficient, in statistical terms, to use fewer grains for the induced group and more in the spontaneous group. For practical reasons, though, it is hard to achieve the optimum statistical choice of numbers of grains in each group. A common strategy, therefore, is to aim for equal numbers of grains in the two groups and to try to choose the neutron fluence to give similar track densities in the two groups. This is more easily said than done, particularly as the area A is dictated by the sizes of the smaller grains.

Table 4.1 shows some spontaneous and induced track counts determined by the population method for a sample of apatite 92-1 from the inter-laboratory comparison reported in Miller *et al.* (1993). In Chapter 3 a sample of this same apatite was dated by the external detector method (Table 3.1). There are 52 grains in the spontaneous group and 48 in the induced group. For each grain, a track count is made over the same area A of surface, in this case $A = 65 \times 10^{-6}$ cm^2. In total there are 1850 spontaneous tracks and 556 induced tracks. The induced track counts are quite small, suggesting that a higher neutron fluence would have been preferable.

For this experiment, the analyst did not use the zeta-calibration approach. However, by back-calculation from the results that were given, the equivalent calibration factors are $\hat{\zeta} = 351$ and $\hat{\rho}_d = 0.14$ in the usual units so that $\hat{\zeta}\hat{\rho}_d = 49.14$ Ma. We will use these values for purposes of illustration and suppose also that $N_d = 3000$ and se$(\hat{\zeta}) = 6$, which are typical values.

Table 4.2 presents some unpublished data obtained using the population method for an apatite sample called DB2 from South Africa. There are 1488 spontaneous tracks counted in 93 grains, and 7885 induced tracks in 120 grains. Again, equal counting areas were used for all grains. The induced track counts here are a good size.

Table 4.2 *Spontaneous and induced track counts determined by the population method for apatite DB2, South Africa.*

Number of fission tracks counted per grain

Spontaneous									
17	8	6	22	16	16	20	19	24	3
21	15	4	39	20	17	10	13	8	20
23	28	17	31	28	8	18	6	10	23
30	4	10	9	34	6	40	9	17	2
31	12	13	10	16	17	9	9	13	14
26	11	9	19	9	14	27	19	7	27
7	35	33	8	11	34	27	9	15	30
11	22	5	4	3	8	14	17	8	5
34	34	10	6	5	16	15	3	16	22
9	15	14							

Induced									
40	56	104	38	36	70	174	148	104	48
56	117	92	146	36	82	81	65	33	38
88	50	64	42	28	51	138	118	51	75
70	44	62	44	28	31	28	39	64	75
93	73	34	83	136	52	55	133	73	24
88	28	28	35	59	34	13	156	30	28
173	16	26	27	18	109	34	41	85	21
163	99	58	27	31	136	36	32	16	106
165	238	16	44	56	51	55	34	14	134
46	134	36	109	79	28	44	31	99	105
36	20	111	27	58	80	97	32	14	34
27	66	51	142	41	68	48	51	57	44

Equal (unspecified) counting areas were used; $\hat{\zeta} = 355$, $se(\hat{\zeta}) = 8$, both in Ma/10^6; $N_d = 5000$ tracks; $\hat{\rho}_d = 1.72 \times 10^6$ tracks/cm^2.

4.2 Theoretical and observed track densities

Suppose there are n grains in the spontaneous group and m in the induced group, and that tracks are counted over the same area A for every grain. Let N_{ij} denote the number of tracks counted for grain j in the induced group. From the line segment model, N_{ij} will be from a Poisson distribution with mean $A\rho_{ij}$, where

$$\rho_{ij} = \tfrac{1}{2}\tau_{ij}\mu_i$$

and where μ_i is the mean length of induced tracks and τ_{ij} is the expected number of fissioned ^{235}U atoms per unit volume in grain j, c.f. equation (2.13). This ρ_{ij} is the theoretical track density for that grain. Since all of the induced tracks were created together and have not been heated, it is reasonable to assume that the mean induced track length μ_i is the same for each grain. But the concentrations τ_{ij} will typically differ between grains. Hence so will the ρ_{ij} and we need to describe this variation statistically.

Assume that the concentrations $\tau_{i1}, \tau_{i2}. \ldots \tau_{im}$ are a random sample from a population with mean $\mu_{\tau i}$, say, and standard deviation $\sigma_{\tau i}$. Let α_i denote the *coefficient of variation* of this population, viz.,

$$\alpha_i = \sigma_{\tau i}/\mu_{\tau i}. \tag{4.1}$$

The parameter α_i is important. It measures the dispersion of these uranium concentrations, relative to their mean, over different grains in the relevant population. (Formally, a value in this population is the number of ^{235}U atoms per unit volume in a given grain that would be expected to fission if that grain were heated and irradiated.) Moreover, because ρ_{ij} is proportional to τ_{ij}, the coefficient of variation of the induced track densities is also α_i.

Thus we imagine that the induced track count N_{ij} for grain j is obtained from the following sampling experiment: choose a grain at random from the population, heat and irradiate it, and then count the induced tracks over an area A. It can be shown by standard calculations that the expected number of tracks counted is then

$$E(N_{ij}) = E(A\rho_{ij}) = A\rho_i , \tag{4.2}$$

where

$$\rho_i = \tfrac{1}{2}\mu_{\tau i}\mu_i \tag{4.3}$$

is the average induced track density over all grains. This is given by the same formula as ρ_{ij} but with τ_{ij} replaced by its mean value $\mu_{\tau i}$. Furthermore, the variance of the number of tracks counted in this experiment is

$$var(N_{ij}) = E(A\rho_{ij}) + var(A\rho_{ij}) = A\rho_i + \alpha_i^2 A^2 \rho_i^2 . \tag{4.4}$$

The first term in this formula is the average variance of a Poisson count about its mean value and the second term is the variance of the means of the Poisson counts over the population of grains. Remember that the variance of a Poisson distribution is equal to its mean. If the concentrations τ_{ij} did not vary, i.e., if $\alpha_i = 0$, then in this experiment, N_{ij} would still have a Poisson distribution with mean and variance both equal to $A\rho_i$.

A similar argument applies to the counts N_{sj} for the n grains in the spontaneous group. For grain j, N_{sj} is from a Poisson distribution with mean $A\rho_{sj}$ where

$$\rho_{sj} = \tfrac{1}{2}\tau_{sj}\mu_{sj}$$

and where μ_{sj} is the mean length (in general, the equivalent isotropic length) of spontaneous tracks in grain j. In this case, τ_{sj} is the expected number of fissioned ^{238}U atoms per unit volume in grain j. Here we also need to allow for the possibility that μ_{sj} might vary between grains. Now assume that the ρ_{sj} vary about a mean ρ_s with standard deviation $\sigma_{\rho s}$, so their coefficient of variation is

$$\alpha_s = \sigma_{\rho s}/\rho_s . \tag{4.5}$$

Thus in the experiment: choose a grain at random from the population and count spontaneous tracks over the area A, the expected number of tracks counted is

$$E(N_{sj}) = E(A\rho_{sj}) = A\rho_s \tag{4.6}$$

and the variance of the number of tracks counted is

$$var(N_{sj}) = E(A\rho_{sj}) + var(A\rho_{sj}) = A\rho_s + \alpha_s^2 A^2 \rho_s^2 . \tag{4.7}$$

For completeness we may write the mean spontaneous track density ρ_s as

$$\rho_s = \tfrac{1}{2}\mu_{Ts}\mu_s, \tag{4.8}$$

where μ_{Ts} and μ_s are the mean values of τ_{sj} and μ_{sj} over all grains in the population.

Now, from equation (3.3), the expected number of ^{238}U atoms that will have fissioned is directly proportional to the corresponding number of ^{235}U atoms that might have fissioned, had that grain been irradiated. If spontaneous fission has been taking place for the same time t in all grains, and if the mean length of spontanous tracks is the same in all grains, then it follows that α_s must equal α_i — that is, the coefficients of variation of the distributions of ρ_{ij} and ρ_{sj} must be the same. In this case, we may write $\alpha_s = \alpha_i = \alpha$, say. Note that allocating grains randomly to each group helps to ensure that the distribution of uranium concnetrations is the same for both groups.

However, if tracks have been forming over different times t, or if μ_{sj} varies between grains (e.g., if grains have experienced different amounts of heat), then α_s will be *greater* than α_i. This property helps us to diagnose whether the true fission track ages vary between grains.

This construction, where a count is obtained from a distribution whose mean is itself randomly sampled from a population, is familiar in statistical modelling. The underlying distribution of means (here the distribution of the uranium concentrations or the distribution of spontaneous track densities) is called a *mixing* distribution. In the present case the counts then have a "mixture Poisson distribution". The parameter α is the coefficient of variation of the mixing distribution, so when we need to give it a name we will call it the "mixing coefficient of variation" or MCV for short.

Equations (4.8) and (4.3) represent *theoretical* track densities as discussed in Section 3.3. The corresponding *estimated* track densities are

$$\hat{\rho}_s = \frac{N_s}{nA} = \frac{\bar{n}_s}{A} \quad \text{and} \quad \hat{\rho}_i = \frac{N_i}{mA} = \frac{\bar{n}_i}{A}, \tag{4.9}$$

where N_s and N_i are the total numbers of spontaneous and induced tracks, respectively, summed over all grains in each group, and $\bar{n}_s$ and $\bar{n}_i$ are the sample mean numbers of tracks per grain. Using (4.7) and (4.4) it can be shown that the relative standard errors of these estimates are

$$\frac{se(\hat{\rho}_s)}{\rho_s} = \left(\frac{1}{nA\rho_s} + \frac{\alpha_s^2}{n}\right)^{\frac{1}{2}} \approx \left(\frac{1}{N_s} + \frac{\alpha_s^2}{n}\right)^{\frac{1}{2}}$$

and

$$\frac{se(\hat{\rho}_i)}{\rho_i} = \left(\frac{1}{mA\rho_i} + \frac{\alpha_i^2}{m}\right)^{\frac{1}{2}} \approx \left(\frac{1}{N_i} + \frac{\alpha_i^2}{m}\right)^{\frac{1}{2}}.$$

Finally, the situation when every grain has the same amount of uranium per unit volume corresponds to $\alpha_s = \alpha_i = 0$. Then all of the above formulae reduce to the same as those in Section 3.3 for the external detector method.

4.3 An estimate of the uranium dispersion

It is important to estimate the parameters α_i and α_s. It is also important to check whether they are the same. We discuss this latter point in Section 4.7. If there is evidence that α_i and α_s differ, this would suggest that the grains were not allocated randomly to the two groups or that the grains had different true fission track ages or that there is some experimental effect distorting the data. Any of these might cast doubt on the validity of the age estimate.

A simple estimate of α_i, based on the method of moments, is

$$\hat{\alpha}_i = \sqrt{\frac{m^2 s_i^2}{N_i^2} - \frac{m}{N_i}} = \sqrt{\frac{s_i^2}{\bar{n}_i^2} - \frac{1}{\bar{n}_i}}, \qquad (4.10)$$

where $\bar{n}_i$ and s_i are, respectively, the mean and standard deviation of the m induced track counts. If there is negligible uranium dispersion, or if the counts are small, it is possible that the term inside the square root is negative. In this case $\hat{\alpha}_i$ is taken to be zero. A similar calculation using the spontaneous track counts gives an estimate of $\hat{\alpha}_s$. Other estimates are possible. For example, if one were prepared to specify a parametric distribution for the uranium concentrations, one could then calculate maximum likelihood estimates.

For the apatite 92-1 (Table 4.1), the mean number of induced tracks per grain is $556/48 = 11.6$ and the standard deviation is $s_i = 3.37$. From (4.10), $s_i^2/\bar{n}_i^2 - 1/\bar{n}_i = -0.0017$, so we set $\hat{\alpha}_i = 0$. This suggests that there is no evidence of variation in uranium concentrations between grains. It should be noted, though, that the track counts are quite small which makes it hard to detect moderate heterogeneity. A similar calculation to (4.10) using the spontaneous track counts gives $\hat{\alpha}_s = 0.230$, representing a 23% relative dispersion between grains over and above that due to Poisson variation. This may suggest that the true fission track ages vary between grains. We return to this point in Section 4.7.

For apatite B2 (Table 4.2), the reader may verify that $\hat{\alpha}_i = 0.648$ (i.e., about 65%), suggesting that the uranium dispersion is substantial. Furthermore, from the spontaneous track counts, $\hat{\alpha}_s = 0.534$, which is comparable with $\hat{\alpha}_i$ — indeed is slightly less — suggesting that there is no evidence of variation in fission track ages between grains.

4.4 A uranium homogeneity test

Sometimes it may be reasonable to suppose a priori that there is no uranium variation (i.e., $\alpha_i = 0$) — for example if the grains were crushed from a single large un-zoned grain. A formal test of this hypothesis uses the Poisson index of dispersion (introduced in Section 2.7) calculated as

$$\chi^2_{\text{stat}} = \sum_{j=1}^{m} \frac{(N_{ij} - N_i/m)^2}{N_i/m} = \frac{(m-1)s_i^2}{\bar{n}_i}. \qquad (4.11)$$

The p-value is determined from the χ^2-distribution with $m - 1$ degrees of freedom. A significantly large value of χ^2_{stat} in (4.11) indicates evidence of uranium heterogeneity.

For apatite 92-1 (Table 4.1), the estimate of α_i is 0, indicating that there is no uranium variation. The χ^2 statistic is 46.1 with 47 degrees of freedom and the p-value is 0.51, confirming formally that the induced counts are consistent with Poisson variation. For apatite DB2 (Table 4.2), the estimate of α_i is 0.648 suggesting that there is considerable over-dispersion. This is confirmed by the χ^2 statistic, which is 3398 with 199 degrees of freedom and a p-value of 0.00 to two decimal places.

4.5 Estimates of fission track age

4.5.1 The fission track age equation

A parallel development to that given in Section 3.5 leads to the following fission track age equation for the population method:

$$t = \frac{1}{\lambda} \log \left(1 + \lambda \zeta \rho_d \frac{\rho_s}{\rho_i} \right),$$
(4.12)

where ρ_s and ρ_i are given by (4.8) and (4.3) and the other terms are as in (3.10). This equation differs from (3.10) in that there is no factor $\frac{1}{2}$ multiplying ρ_s/ρ_i. This is because here grains are irradiated before they are polished and etched to expose internal surfaces, so that the induced tracks revealed may emanate from ^{235}U atoms on either side of the polished surface. Hence the factor $\frac{1}{4}$ in equation (3.4) becomes $\frac{1}{2}$ in (4.3), leading to (4.12).

4.5.2 Estimating a common age

It is sometimes thought that for the population method to provide a valid age estimate, the uranium concentration needs to be the same in each grain. In fact, as the preceding argument shows, a valid estimate can be obtained if the uranium concentrations vary, provided that they can be regarded as a random sample from a single population, regardless of which group the grains are in. For this reason it is good practice to divide the grains into the two groups using a randomisation device. However, the age estimate will be more precise if the uranium concentrations do not vary.

Substituting the estimates of the track densities given by (4.9) into equation (4.12) gives the age estimate

$$\hat{t} = \frac{1}{\lambda} \log \left(1 + \lambda \zeta \hat{\rho}_d \frac{N_s}{n} \frac{m}{N_i} \right),$$
(4.13)

where $\lambda = 1.55125 \times 10^{-4}$ Ma^{-1} as usual. The factor A cancels out in the calculation of $\hat{\rho}_s/\hat{\rho}_i$, which becomes the ratio of the sample mean numbers of tracks per grain. The approximate relative standard error of $\hat{t}$ is then

$$\frac{se(\hat{t})}{\hat{t}} \approx \left(\frac{1}{N_s} + \frac{1}{N_i} + \hat{\alpha}^2 \left[\frac{1}{n} + \frac{1}{m} \right] + \frac{1}{N_d} + \left(\frac{se(\hat{\zeta})}{\hat{\zeta}} \right)^2 \right)^{\frac{1}{2}}, \qquad (4.14)$$

which takes into account error in estimating ζ and ρ_d as well as ρ_s/ρ_i. Compared with equation (3.13), this formula has an additional term arising from the uranium dispersion α, which can be estimated by $\hat{\alpha} = \hat{\alpha}_i$ from (4.10). Although one could estimate a common α using both the spontaneous and induced track counts, it is preferable to use the induced track counts only because the spontaneous track counts may contain additional sources of variation. The remarks made below equation (3.13) regarding the different components of error apply here too.

For example, for apatite 92-1 (Table 4.3), the age estimate is

$$\hat{t} = \frac{1}{\lambda} \log \left(1 + \lambda \times 351 \times 0.14 \times \frac{1850}{52} \times \frac{48}{556} \right) = 149.2 \text{ Ma}$$

and, using the estimate $\hat{\alpha} = 0$ obtained earlier, the relative standard error is

$$\frac{se(\hat{t})}{\hat{t}} \approx \left(\frac{1}{1850} + \frac{1}{556} + 0.0^2 \left[\frac{1}{52} + \frac{1}{48} \right] + \frac{1}{3000} + \left(\frac{6}{351} \right)^2 \right)^{\frac{1}{2}} = 0.0544 \,.$$

The standard error of $\hat{t}$ is therefore $149.2 \times 0.0544 = 8.1$ Ma. The age estimate differs from the independent age by $(149.2 - 162)/8.1 = -1.6$ standard errors and therefore is reasonably consistent with it. In this example, there is no evidence of uranium variation in the induced group, and therefore the relative standard error is small. However, the induced track density is low, which makes it difficult to detect any moderate heterogeneity.

For apatite DB2 (Table 4.4), the age estimate and its relative standard error are

$$\hat{t} = \frac{1}{\lambda} \log \left(1 + \lambda \times 355 \times 1.72 \times \frac{1488}{93} \times \frac{120}{7885} \right) = 147.0 \text{ Ma}$$

and, using $\hat{\alpha} = \hat{\alpha}_i = 0.648$ from Section 4.3,

$$\frac{se(\hat{t})}{\hat{t}} \approx \left(\frac{1}{1488} + \frac{1}{7885} + 0.648^2 \left[\frac{1}{93} + \frac{1}{120} \right] + \frac{1}{5000} + \left(\frac{8}{355} \right)^2 \right)^{\frac{1}{2}}$$

$$= 0.0976 \,.$$

The standard error of $\hat{t}$ is therefore $147.0 \times 0.0976 = 14.3$ Ma. In this example, the spontaneous track counts are low, but the neutron fluence has been chosen to give reasonably large induced track counts, and hence a reasonably precise age estimate, given the large amount of uranium variation. In general, the precision is dominated by the largest term in the above expression. Here, the largest term is $0.648^2 (1/93 + 1/120) = 0.008$, so precision might be improved by increasing the number of grains. But because $1/7855 = 0.0001$ is already much smaller, there would be little point in increasing the neutron fluence to get larger induced track counts.

4.5.3 Estimating single grain ages

A separate fission track age for each grain can be calculated by using the single grain count N_{sj} and $n = 1$ in equations (4.13) and (4.14). For example, for the first grain in the spontaneous group in apatite DB2 (Table 4.2), 17 tracks were counted. So for this grain

$$\hat{t} = \frac{1}{\lambda} \log \left(1 + \lambda \times 355 \times 1.72 \times \frac{17}{1} \times \frac{120}{7885} \right) = 156.1 \text{ Ma},$$

with relative standard error

$$\frac{se(\hat{t})}{\hat{t}} \approx \left(\frac{1}{17} + 0.648^2 \right)^{\frac{1}{2}} = 0.692.$$

This relative standard error uses only the terms $1/N_{sj}$ and $\hat{a}^2/n$ in equation (4.14) so does not account for error in estimating ρ_i, ρ_d or ζ. This is the appropriate relative standard error for comparing age estimates of single grains within the same sample. The same estimates of ρ_i, ρ_d and ζ are used for each grain and therefore the errors in these estimates do not contribute to comparisons between them. In any case, for a single grain (4.14) is usually dominated by these two terms because $n = 1$ and hence N_s may not be very large either. In fact, with the other terms added the relative standard error here is 0.694, which is practically the same. Note that the precision of a single grain age estimate is necessarily low; here the relative standard error is 69%.

4.6 Summarising and inspecting the data

For data obtained by the population method, useful summary statistics include the total numbers of grains (m and n), the total numbers of tracks (N_s and N_i) the sample means and standard deviations of the numbers of tracks per grain, and the uranium dispersion measures (i.e., the mixing coefficients of variation) $\hat{a}_i$ and $\hat{a}_s$. These statistics are given in Table 4.3 for the apatite 92-1 sample in Table 4.1.

Table 4.3 *Summary statistics for apatite 92-1 (Table 4.1).*

	Number of grains	Number of tracks	Mean tracks per grain	s.d. tracks per grain	MCV $\hat{a}$
Spontaneous	52	1850	35.6	10.13	0.230
Induced	48	556	11.6	3.37	0.000

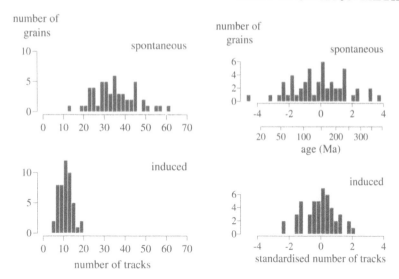

Figure 4.1 *Left panels: histograms of the numbers of spontaneous and induced tracks for apatite 91-1 (Table 4.1). Right panels: histograms of the standardised numbers of tracks $z - z_0$ where z is given by (4.15). These should scatter with unit standard deviation, mostly between -2 and $+2$. Comparison of the two right panels suggests that the spontaneous track counts are over-dispersed.*

There are usually too many grains to show single grain ages in a table, so it is convenient to look at them, and at the raw counts, graphically. Although the counts are discrete, they usually vary enough to present them sensibly in histograms. The left panels in Figure 4.1 show histograms of the raw counts for each group with respect to the same scale. Clearly the spontaneous track counts are both larger and more variable than the induced ones. But it is not possible to infer from this graph that they are over-dispersed because, even when $\alpha = 0$, the variance increases with the mean — i.e., counts from a Poisson distribution vary more when the mean is higher.

In the right-hand panels in Figure 4.1 the two samples are plotted in such a way as to correct for this, so that the greater variation seen here in the spontaneous group really does indicate that α_s is greater than α_i. These panels show histograms of *standardised* values. For the induced group the standardised numbers are $z_{ij} - z_{i0}$ where

$$z_{ij} = 2\sqrt{N_{ij}} \qquad (4.15)$$

and z_{i0} is a convenient central value. Here $z_{i0} = 2\sqrt{\bar{n}_i} = 2\sqrt{11.6} = 6.81$, i.e., obtained by using the mean count instead of an individual count in (4.15). For the spontaneous group standardised values $z_{sj} - z_{s0}$ are similarly defined in terms of N_{sj} and $\bar{n}_s$. This method uses the fact that the square root transformation is a *variance-stabilising* transformation for the Poisson distribution. If

Table 4.4 *Summary statistics for apatite DB2 (Table 4.2).*

	Number of grains	Number of tracks	Mean tracks per grain	s.d. tracks per grain	MCV $\hat{\alpha}$
Spontaneous	93	1488	16.0	9.43	0.534
Induced	120	7885	65.7	43.32	0.648

N has a Poisson distribution with mean μ, then (provided μ is not too small) the variance of $2\sqrt{N}$ is approximately 1, regardless of the value of μ. We are using the square root transformation here because the induced track counts vary consistently with a Poisson distribution (and in fact $\hat{\alpha}_i = 0$).

In the right-hand panels of Figure 4.1 the standardised numbers in the induced group are nearly all between -2 and $+2$ as one would expect, and their standard deviation is 1.04. Because we have used the square root transformation, this is consistent with Poisson variation and in agreement with the dispersion test in Section 4.4. But those in the spontaneous group (i.e., $z_{sj} - z_{s0}$) clearly vary more, suggesting that the true single grain ages may vary. The upper panel is equivalent to a histogram of single grain ages, so an age scale has been added. This is of course a non-linear scale. The formula for matching the t scale to the $z - z_{s0}$ scale is

$$z - z_{s0} = 2\sqrt{\left(\frac{e^{\lambda t} - 1}{\lambda}\right)\frac{\bar{n}_i}{\hat{\zeta}\hat{\rho}_d}} - 2\sqrt{\bar{n}_s}.$$

Here $z_{s0} = 2\sqrt{\bar{n}_s}$ corresponds to t being the common age estimate of 149.2 Ma.

For apatite DB2 (Table 4.2) the summary statistics and histograms are given in Table 4.4 and Figure 4.2. The mean number of tracks per grain is larger for the induced group, and so is the standard deviation, as can be seen in the table and in the left panels of Figure 4.2. The values of $\hat{\alpha}_i$ and $\hat{\alpha}_s$ are both quite large, but not very different, suggesting that there is substantial variation in uranium concentration between grains but little or no evidence of any variation in ages.

Because there is substantial extra-Poisson variation in the induced track counts ($\hat{\alpha}_i = 0.648$) the square root transformation will not do to stabilise their variance. A transformation that may be used in this case, where the mean and variance are related by (4.2) and (4.4), is the inverse hyperbolic sine transformation. Here the standardised values are $z_{ij} - z_{i0}$ with z_{ij} now given by

$$z_{ij} = 2\alpha^{-1}\operatorname{arcsinh}\left(\alpha\sqrt{N_{ij}}\right) \tag{4.16}$$

$$= 2\alpha^{-1}\log\left(\alpha\sqrt{N_{ij}} + \sqrt{\alpha^2 N_{ij} + 1}\right), \tag{4.17}$$

where $\operatorname{arcsinh}(x) = \log(x + \sqrt{x^2 + 1})$. Again z_{i0} has been chosen by substituting the mean track count $\bar{n}_i$ in place of the individual track count N_{ij} in

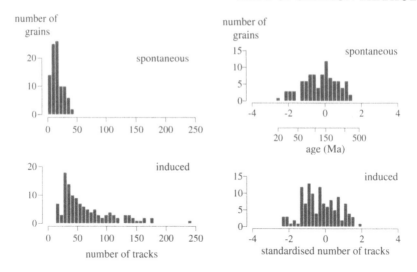

Figure 4.2 *Left panels: histograms of numbers of spontaneous and induced tracks from apatite DB2 (Table 4.2). Right panels: histograms of standardised values $z - z_0$ from (4.17). Both groups are overdispersed with respect to Poisson variation but the dispersion is approximately the same. This is consistent with different uranium concentrations but the same fission track age for all grains.*

this expression. Standardised values for the spontaneous group are defined similarly using N_{sj} and $\bar{n}_s$. In this case, the transformation depends on the parameter α and it is crucial to use the *same* value for both groups, obtained from the *induced* track counts only. Here $\hat{\alpha} = \hat{\alpha}_i = 0.648$.

In Figure 4.2 the standardised values for the induced group should scatter about 0 with unit standard deviation approximately, which they do. So do those for the spontaneous group also, suggesting that the single grain ages are consistent with a common true value. Again, an age scale has been added. Here the formula for matching the t scale to the $z - z_{s0}$ scale is

$$ z - z_{s0} = \frac{2}{\alpha} \operatorname{arcsinh} \left(\alpha \sqrt{\left(\frac{e^{\lambda t} - 1}{\lambda} \right) \frac{\bar{n}_i}{\hat{\zeta} \hat{\rho}_d}} \right) - \frac{2}{\alpha} \operatorname{arcsinh} \left(\alpha \sqrt{\bar{n}_s} \right), $$

where again z_{s0} corresponds to the common age estimate.

Why do we need radial plots for the external detector method but not for the population method? Compare Figure 4.2 with Figure 3.2, for example. In both cases there is a relation between the variance and the mean of the estimated single grain ages, where the variance increases with the mean. For the population method, this relation can be removed by a suitable transformation, because every spontaneous track density is compared to the same induced track density $\hat{\rho}_i = \bar{n}_i / A$. But for the external detector method each spontaneous track density has its own induced track density that differs between grains. This is inherent in the method. In fact, these individual track densities give us extra information. Not only do they improve precision but they also allow more scope for modelling samples with different ages, as we will see in the next two chapters.

A useful table of variance stabilising transformations can be found in the textbook by Johnson and Leone (1964, p. 56). The inverse hyperbolic sine transformation is applicable when a count N is from a distribution with mean μ and variance of the form $\mu + \alpha^2 \mu^2$. Some refinements to improve the transformations when the counts are small are given by Anscombe (1948, 1950) — for example, adding $\frac{3}{8}$ to each count is a good rule. Such transformations are not used much these days for statistical modelling, but they are still very useful for graphical analysis.

4.7 Age homogeneity test

If all grains have the same true fission track then the dispersion parameter α must be the same for both spontaneous and induced track counts, i.e., $\alpha_s = \alpha_i$.

When equal counting areas are used for every grain, a simple method to test this hypothesis is to use the transformed values z_{ij} defined by equation (4.17) and the corresponding values z_{sj} for the spontaneous group. For these values we simply test the hypothesis that the two groups have the same variance using a standard F-test (Snedecor and Cochran, 1980, Section 6.12; Armitage et al., 2002, p. 150). To carry out this test, calculate the ratio of sample variances (spontaneous/induced) of the zs and determine the p-value from the upper tail of the F-distribution with $n - 1$ and $m - 1$ degrees of freedom. This is a one-sided test because in principle α_s should not be less than α_i.

For apatite DB2, $\hat{\alpha}_s < \hat{\alpha}_i$, so there is no need to carry out a formal test. However, we will do so for illustrative purposes. The sample standard deviation of the z_{ij} values is 0.2996, and likewise the sample standard deviation of the z_{sj} values is 0.3140. So the F-ratio is

$$F_{stat} = \frac{0.2996^2}{0.3140^2} = 0.911 ,$$

with degrees of freedom 92 and 119. The p-value is 0.679, confirming that the data are consistent with $\alpha_s = \alpha_i$.

For apatite 92-1, $\hat{\alpha} = 0$. In this case we use the square root transformed values from (4.15). The sample standard deviations of these z-values are 0.8538 and 0.5021. The F-ratio is

$$F_{stat} = \frac{0.8538^2}{0.5021^2} = 2.89 ,$$

with degrees of freedom 51 and 47. The p-value is 0.00017, which is strong evidence that α_s is greater than α_i. This confirms the visual impression from Figure 4.1.

4.8 A measure of dispersion of true fission track ages

Suppose that the true fission track ages vary between grains. For example, tracks may have been shortened by heat but by differing amounts in different grains because of their differing chemical compositions. Let α_t be the coefficient of variation of the distribution of true ages. Then it can be shown that, approximately,

$$\alpha_s^2 = \alpha_t^2 + \alpha_i^2 .$$

Hence, when $\hat{\alpha}_s > \hat{\alpha}_i$, we may estimate α_t by

$$\hat{\alpha}_t = \sqrt{\hat{\alpha}_s^2 - \hat{\alpha}_i^2}$$

and when $\hat{\alpha}_s \leq \hat{\alpha}_i$, let $\hat{\alpha}_t = 0$. The quantity α_t has a similar interpretation to the age dispersion σ discussed in Section 3.9 in relation to the external detector method.

It can be shown (Anscombe, 1950) that the standard error of the estimate of α_i^2 given by (4.10) is approximately

$$\text{se}(\hat{\alpha}_i^2) \approx \frac{s_i^2}{\bar{n}_i^2} \sqrt{\frac{2}{m} \left(1 + \hat{\alpha}_i^2 \right)} . \tag{4.18}$$

Similarly

$$\text{se}(\hat{\alpha}_s^2) \approx \frac{s_s^2}{\bar{n}_s^2} \sqrt{\frac{2}{n} \left(1 + \hat{\alpha}_s^2 \right)} . \tag{4.19}$$

Hence it can be shown that an approximate standard error for $\hat{\alpha}_t$ is

$$\text{se}(\hat{\alpha}_t) \approx \left(\frac{[\text{se}(\hat{\alpha}_i^2)]^2 + [\text{se}(\hat{\alpha}_s^2)]^2}{4 \, \hat{\alpha}_t^2} \right)^{\frac{1}{2}} .$$

For apatite 92-1 (see Tables 4.4 and 4.6) the estimated coefficient of variation of the true ages is

$$\hat{\alpha}_t = \sqrt{0.230^2 - 0.000^2} = 0.230 .$$

Using equations (4.18) and (4.19) the approximate standard error of this estimate is

$$\text{se}(\hat{\alpha}_t) \approx \left(\frac{[0.0173]^2 + [0.0163]^2}{4 \times 0.230^2} \right)^{\frac{1}{2}} = 0.052 .$$

This standard error is small compared with the estimate 0.230, confirming that the dispersion is greater than 0, and giving a rough idea of its magnitude.

In the previous section, we found that for apatite DB2, there is no evidence of age dispersion, but for apatite 92-1, there was strong evidence. This is perhaps surprising, as apatite 92-1 is a tentative age standard, but Miller *et al.* (1993) noted that this sample from a volcanic ash did contain some older grains.

4.9 Counts over unequal areas

Most of the preceding analyses can be adapted quite simply to the situation where numbers of tracks are counted over areas that differ from grain to grain. Specifically, let N_{ij} denote the number of tracks counted over an area A_{ij} for grain j in the induced group, with corresponding definitions for N_{sj} and A_{sj} for grains in the spontaneous group.

If uranium concentrations do not vary between grains then it is immaterial what areas are used in which grains. We may regard the total count in each group as coming from just one "super-grain" on which a Poisson count is made — totals of N_s spontaneous tracks and N_i induced tracks, say. The estimated track densities are

$$\hat{\rho}_s = \frac{N_s}{A_s} \quad \text{and} \quad \hat{\rho}_i = \frac{N_i}{A_i}, \tag{4.20}$$

where A_s and A_i are the total areas over the n and m grains, respectively. These estimates can also be viewed as *weighted* averages of the estimated track densities for each grain with weights proportional to the areas. In the absence of uranium variation, their relative standard errors are

$$\frac{\mathrm{se}(\hat{\rho}_s)}{\rho_s} \approx \left(\frac{1}{N_s}\right)^{\frac{1}{2}} \quad \text{and} \quad \frac{\mathrm{se}(\hat{\rho}_i)}{\rho_i} \approx \left(\frac{1}{N_i}\right)^{\frac{1}{2}}.$$

These $\hat{\rho}_s$ and $\hat{\rho}_i$ may then be used in the fission track age equation (4.12) and their two relative standard errors may be combined as usual, along with the other relevant components of error, as the square root of the sum of their squares.

When the uranium concentrations do vary, we may still follow the argument in Section 4.2, where we regard τ_{ij} as a random value from a population with coefficient of variation α_i given by equation (4.1). But we also need to make some assumption about the relationship (if any) between τ_{ij} and A_{ij}, the area chosen in that grain. We will make the simplest assumption, namely that they are independent, though this is not the only possibility.

Now imagine the experiment: choose a grain at random from the population, heat and irradiate it, and then count the induced tracks over an area A_{ij}. Under the above assumptions, the count N_{ij} in this experiment, with A_{ij} fixed, will have expected value and variance given by

$$A_{ij}\rho_i \quad \text{and} \quad A_{ij}\rho_i + \alpha_i^2 A_{ij}^2 \rho_i^2,$$

where ρ_i is the population mean induced track density over all grains. By a corresponding argument, a spontaneous track count N_{sj} will have expected value and variance given by

$$A_{sj}\rho_s \quad \text{and} \quad A_{sj}\rho_s + \alpha_s^2 A_{sj}^2 \rho_s^2,$$

where ρ_s is the population mean spontaneous track density. Furthermore, if all grains have the same true fission track age then $\alpha_s = \alpha_i = \alpha$ as before.

Under this model, the estimates of ρ_s and ρ_i given by (4.20) are unbiased and they have approximate relative standard errors given by

$$\frac{\text{se}(\hat{\rho}_s)}{\rho_s} \approx \left(\frac{1}{N_s} + \frac{\alpha^2 B_s}{A_s^2}\right)^{\frac{1}{2}} \quad \text{and} \quad \frac{\text{se}(\hat{\rho}_i)}{\rho_i} \approx \left(\frac{1}{N_i} + \frac{\alpha^2 B_i}{A_i^2}\right)^{\frac{1}{2}},$$

where $B_s = \sum_{j=1}^{n} A_{sj}^2$ and $B_i = \sum_{j=1}^{m} A_{ij}^2$ are the sums of the *squares* of the areas used in each group.

As before, we estimate α from the induced track counts. A fairly simple estimate is obtained by equating the weighted variance of the estimated induced track densities to its expected value and then substituting $\hat{\rho}_i$ in place of ρ_i. This leads to

$$\hat{\alpha} = \hat{\alpha}_i = \sqrt{\frac{s_{wi}^2 - \hat{\rho}_i}{\hat{\rho}_i^2(\bar{a}_i - s_{ai}^2/m\bar{a}_i)}}, \tag{4.21}$$

where $\bar{a}_i$ and s_{ai} are the mean and standard deviation of the m areas A_{ij} and

$$s_{wi}^2 = \frac{1}{m-1} \sum_{j=1}^{m} A_{ij} \left(\frac{N_{ij}}{A_{ij}} - \frac{N_i}{A_i}\right)^2.$$

If the expression inside the square root in (4.21) is negative, then $\hat{\alpha}$ is set to zero as before. Equation (4.21) reduces to (4.10) when all areas are the same.

4.10 A protocol for data analysis

The statistical analyses in the preceding sections are presented in a convenient order largely dictated by the theory. In practice, one is usually confronted with some data that need to be interpreted, and questions may arise in a different order. As a general rule it is best to tabulate, plot and inspect data before applying more formal methods. In any case, a suggested protocol would include the following steps, in some order:

1. Tabulation of single grain data and summary statistics.
2. Graphs of counts and standardised counts.
3. Assessment of the amount of uranium variation.
4. Calculation of a common age and its precision.
5. Age homogeneity test.
6. Calculation of age dispersion.

Again, further analysis would depend on the geological context and on what is learnt after these steps.

4.11 Discussion

The main perceived advantage of the population method is that both spontaneous and induced fission tracks are counted in the crystal, thus avoiding possible technical problems with the external detector method, such as poor contact between the grain and the external detector, as discussed in Section 3.12. When there is little or no uranium heterogeneity, the method can give a reasonably precise estimate for a sample of grains with a common age, and can also give single grain age estimates.

But the population method has quite serious limitations for inferences compared with the external detector method. Firstly, there is an extra source of variation contributing to the age estimate: variation in uranium content between grains. This can be substantial, as values of α greater than 0.20, and even greater than 0.50, are common, so the precision of the age estimate is reduced. By the same token it is harder to assess the extent of any spread in true ages and to estimate single grain ages — and therefore it is harder to unravel components in samples with mixed ages.

Secondly, because the material is split into two groups, there may be a waste of information, particularly in samples with a low yield. With the external detector method we could in principle get both spontaneous and induced track counts from every grain.

Thirdly, many grains may be needed to obtain useful results. It is hard to give an unqualified statement of how many — it depends on α among other things — but a guide would be to have both n and m greater than 50 and preferably greater than 100.

Fourthly, the population method is usually conducted by counting equal areas in all grains. This restriction could result in some loss of information because the common area is dictated by the smallest grains. Unequal areas can be used, but this complicates the analysis. Differing grain sizes presents no problem to the external detector method.

Finally, the method is really only applicable to apatite. In zircon and sphene, for example, crystals accumulate radiation damage other than by spontaneous fission of ^{238}U and heating these grains to remove spontaneous tracks prior to irradiation changes their character. Gleadow (1981) showed that the population method applied to zircon and sphene over-estimated the true age systematically. This is one reason why the population-subtraction method is sometimes used. We summarise the corresponding formulae for this method below.

4.12 The population-subtraction method

In this variant of the population method, grains are again split into two groups, ideally using a suitable randomisation device. One group is mounted, polished, and etched to reveal spontaneous tracks, whereas the other group is first irradiated, and then polished and etched, so the tracks revealed will be a mixture of the spontaneous tracks formed over geological time and the induced tracks formed in the irradiation. The observer, though, is unable to tell which are which and simply counts all tracks in the required area.

This design avoids heating the grains and possibly changing their character. However, it can be quite badly affected by problems associated with uranium heterogeneity or zonation, and may therefore result in ambiguities. There is a further loss of precision because the induced track density is measured indirectly. Statistical analysis parallels that for the population method, so we will just comment briefly on the modified formulae.

The primary use of the population-subtraction method is for dating glass. Typically shards and pumice fragments have a wide range in size. The fossil track density is usually very low, particularly in Quaternary tephras, so the analyst wants to count the maximum possible area of each glass fragment. Therefore a different area is usually counted in each grain. Furthermore, in glasses it is often reasonable to assume that uranium concentration does not vary.

First consider the less usual case where the same area A is counted for each grain. Here we can allow for possible uranium heterogeneity. The counts are now denoted by N_{sj}, for $j = 1, 2, \ldots, n$ and $N_{s+i,j}$, for $j = 1, 2, \ldots, m$, where $N_{s+i,j}$ is the number of tracks counted in the jth grain in the irradiated group. As the notation implies, this will be the sum of the spontaneous and induced tracks in that grain. Thus $N_{s+i,j}$ has expected value $A\rho_s + A\rho_i = A(\rho_s + \rho_i)$.

The track densities ρ_s and ρ_i are therefore estimated by

$$\hat{\rho}_s = \frac{N_s}{A_s} \quad \text{and} \quad \hat{\rho}_i = \frac{N_{s+i}}{A_{s+i}} - \frac{N_s}{A_s}, \tag{4.22}$$

where N_s is the total number of tracks and $A_s = nA$ is the total area for all n grains in the spontaneous group, and N_{s+i} is the total number of tracks and $A_{s+i} = mA$ is the total area for all m grains in the irradiated group.

With these estimated track densities, the estimate of the common age is given by equation (4.13), viz.,

$$\hat{t} = \frac{1}{\lambda} \log \left(1 + \lambda \hat{\zeta} \hat{\rho}_d \frac{\hat{\rho}_s}{\hat{\rho}_i} \right), \tag{4.23}$$

where $\lambda = 1.55125 \times 10^{-4}$ Ma^{-1} as usual to give $\hat{t}$ in Ma. Now the relative standard error is

$$\frac{\text{se}(\hat{t})}{t} \approx \left(\frac{1}{N_s} + \frac{1}{N_{s+i}} + \hat{\alpha}^2 \left[\frac{1}{n} + \frac{1}{m} \right] + \frac{1}{N_d} + \left(\frac{\text{se}(\hat{\zeta})}{\zeta} \right)^2 \right)^{\frac{1}{2}} \left(1 + \frac{\hat{\rho}_s}{\hat{\rho}_i} \right). \tag{4.24}$$

Compared with (4.14) this has been inflated by the factor $1 + \hat{\rho}_s/\hat{\rho}_i$. This expresses the loss in precision compared with the usual population method because the induced track density has been determined indirectly by subtracting $\hat{\rho}_s$ from $\hat{\rho}_{s+i}$. To keep this factor small requires ρ_s to be small compared with ρ_i. For example, if ρ_s/ρ_i is $\frac{1}{4}$, the relative standard error would be inflated by 25%. A high precision is difficult to achieve because if ρ_i is too large, it is impossible to count the tracks; and if ρ_s is small the spontaneous track counts will be small and so precision will be lost.

The estimate of α is given by equation (4.10) but using the counts $N_{s+i,j}$ in the irradiated group. Because these contain some spontaneous tracks also, we do not have a pure estimate of the dispersion of uranium concentrations.

Furthermore, it would be very difficult to carry out an age homogeneity test as described above for the population method.

Now consider the more usual case in which the population-subtraction method is used, namely for dating glasses. Here different areas are typically counted, denoted by A_{sj}, for $j = 1, 2, \ldots, n$ and $A_{s+i,j}$, for $j = 1, 2, \ldots, m$. The total areas are

$$A_s = \sum_{j=1}^{n} A_{sj} \quad \text{and} \quad A_{s+i} = \sum_{j=1}^{m} A_{s+i,j} \, .$$

Furthermore it is assumed that the uranium concentration is constant (i.e., $\alpha = 0$). The estimates of track densities and age are given by (4.22) and (4.23) with these A_s and A_{s+i}. This age estimate then has relative standard error

$$\frac{\text{se}(\hat{t})}{t} \approx \left(\frac{1}{N_s} + \frac{1}{N_{s+i}} + \frac{1}{N_d} + \left(\frac{\text{se}(\hat{\zeta})}{\zeta} \right)^2 \right)^{\frac{1}{2}} \left(1 + \frac{\hat{\rho}_s}{\hat{\rho}_i} \right), \qquad (4.25)$$

which is the same as equation (4.24) but with $\hat{\alpha} = 0$. The remarks following (4.24) also apply here.

As mentioned earlier, the population-subtraction method is often used for dating volcanic glasses. A difficulty with this application is that most natural glasses partially anneal over geological time, even at ambient temperatures on the earth's surface, so that the fission track ages obtained will be under-estimates. To some extent the degree of partial annealing can be assessed from the diameters of the etched tracks. If these tend to be smaller in the non-irradiated group (which contains spontaneous tracks only) then some partial annealing is likely to have occurred. One method of correcting for this is to calibrate the diameter reduction against the expected age reduction (Storzer and Wagner, 1969; Bigazzi et al., 1993; and Sandhu and Westgate, 1995). Another is the "plateau" correction method, where pairs of irradiated and non-irradiated glasses are heated together in a furnace (Storzer and Poupeau, 1973; Burchart et al., 1975; Westgate, 1989). The idea behind this is that the furnace annealing reduces the lengths of the induced tracks down to those of the spontaneous tracks, and then the ratio of their densities will give the correct age. The spontaneous tracks are also shortened, but not as much as the induced tracks, which eventually catch up. This method has been used successfully to date partially annealed glass samples ranging in age from Cretaceous to late Quaternary (Westgate, 1989; Sandhu et al., 1993).

4.13 Remarks

This chapter is about the statistical theory underlying the population method and statistical analyses for routine use. Formulae are given that parallel those for the external detector method in Chapter 3 — to calculate age estimates and their precisions when the grains from a single sample have a common age, and measures of age dispersion when there is a spread in fission track ages.

Alternative formulae could be used in some situations — for example, when all counts are large or if further distributional assumptions are made — but those given here can be used safely in nearly all situations and are generally robust.

Gleadow (1981) argued quite persuasively that the usual population method is only applicable to apatite, amongst the common minerals. However, applying the population method to apatite can also be risky, because the grains may not have a common age, due to their differing thermal sensitivities or to the presence of detrital grains. Somewhat ironically, the ability of apatite to repair alpha recoil damage, which makes it suitable for the population method, is probably linked to its thermal sensitivity, which, along with compositional variation, makes it likely that fission track ages will differ between grains. A variant of the population method, the population-subtraction method, is commonly used for dating glass.

4.14 Bibliographic notes

Descriptions of the population method and its relation to other experimental methods are given in various text books and articles, including Fleischer, Price and Walker (1975), Wagner and Van den haute (1992), Naeser (1979a), Naeser and Naeser (1984), Gleadow (1981) and Dumitru (2000). Some of the statistical analyses here are based on Galbraith (1984).

CHAPTER 5

Discrete mixtures of ages

A standard default model for dating a sample is to assume that all grains have the same *true* fission track age. This is the common age model illustrated in Figure 3.5, which would apply, for example, to apatite grains from a single source experiencing a thermal history such as (*a*) in Figure 1.3. Methods for estimating the common age and its precision, and for checking that the data are consistent with this model, are dealt with in Chapters 3 and 4.

Often there is evidence that the true fission track ages may vary between grains. This raises new questions: why might they vary and what quantities can usefully be calculated? This chapter and the next deal with statistical models for this situation and with methods for estimating parameters of interest. Of course, an appropriate choice of model should reflect the relevant geological context. In contrast to Chapters 3 and 4, therefore, the analyses here are not routine, but rather depend on the geological circumstances.

We concentrate on data obtained by the external detector method, consisting of a matched pair of counts of spontaneous and induced tracks for each of several grains, along with estimates of the dosimeter track density ρ_d and the calibration constant ζ. We will usually suppose that all grains were irradiated together to produce the induced tracks and that all tracks were counted by a single analyst, so that the same estimates of ρ_d and ζ apply to each grain. In this case, estimating the fission track age is equivalent to estimating the ratio ρ_s/ρ_i of theoretical spontaneous and induced track densities. In the common age model, this ratio is the same for each grain. In the models to be considered later we specify how this ratio might vary. Sometimes induced track counts from a sample or suite of samples may have been obtained from more than one irradiation, so that the same ρ_d may not apply to all grains. We consider this in Section 5.8.

The external detector method is particularly efficient for assessing variation in true ages because uranium heterogeneity, which is usually substantial, does not contribute to variation in the observed single grain age estimates. Furthermore, for grains with the same true age, the only notable variation in observed ages comes from the Poisson variation in the track counts, which is usually relatively small and easily measured. Similar models can be developed for dating by the population method. But in this case variation in uranium concentration between grains will contribute, often substantially, to variation in observed single grain ages, and it is much harder to assess variation in true ages over and above this.

With respect to statistical theory, these two chapters use maximum likelihood estimation and associated inference. This is well-established methodology, applicable to a wide class of parametric statistical models, that provides estimates of parameters and their approximate standard errors along with methods for testing hypotheses that provide a basis for choosing between models. The definition of the likelihood function and some notes on maximum likelihood inference are given in the Appendix, Section A.10.

5.1 Maximum likelihood estimation of a common age

Before formulating models for varying ages, it is helpful to review the common age model in the context of maximum likelihood inference. It transpires here that the pooled age given by equation (3.12) is the *maximum likelihood* estimate derived from the Poisson distributions for the track counts.

Specifically, let $N_{su}, N_{iu}, u = 1, 2, \ldots, n$ denote spontaneous and induced track counts for n grains, obtained by the external detector method. Suppose that these grains were irradiated together and that tracks were counted by a single analyst, so that the same estimates of ρ_d and ζ apply to each grain. From Section 3.3, N_{su} and N_{iu} have independent Poisson distributions with means $A_u \rho_{su}$ and $A_u \rho_{iu}$. Under the common age model, the ratio of these means is the same for each grain, i.e., $\rho_{su}/\rho_{iu} = \rho_s/\rho_i$ for all u, and estimating the fission track age is equivalent to estimating the ratio ρ_s/ρ_i.

It is advantageous to define the parameter of interest to be the natural log of this ratio, which we denote by β. That is, let

$$\beta = \log\left(\frac{\rho_s}{\rho_i}\right) \tag{5.1}$$

so that $\rho_s/\rho_i = e^\beta$. We wish to estimate β and hence the fission track age by substituting the estimate of $\rho_s/\rho_i = e^\beta$ into equation (3.10).

For convenience, write $\mu_u = A_u \rho_{iu}$. Then the Poisson means are $E(N_{su}) = e^\beta \mu_u$ and $E(N_{iu}) = \mu_u$. The likelihood function is proportional to the joint probability that the counts attain their observed values. This is the product of the separate Poisson probabilities for each count. The log-likelihood function is the natural log of this product (plus an arbitrary constant) and can be shown to be given by

$$L = \sum_{u=1}^{n} \left\{ \beta N_{su} - (1 + e^\beta)\mu_u + (N_{su} + N_{iu}) \log \mu_u \right\} + \text{constant} \tag{5.2}$$

(see the Appendix, Section A.10.1). This is regarded as a function of the $n+1$ parameters $\mu_1, \mu_2, \ldots, \mu_n$ and β.

The maximum likelihood estimates are the values of the parameters for which L is a maximum. These may be found by standard methods, and are given by

$$\hat{\beta} = \log\left(\frac{N_s}{N_i}\right) \tag{5.3}$$

and

$$\hat{\mu}_u = \frac{N_i(N_{su} + N_{iu})}{N_s + N_i} \qquad u = 1, 2, \ldots, n \,,$$

where N_s and N_i are the total numbers of spontaneous and induced tracks summed over all grains. Furthermore, the approximate standard error of $\hat{\beta}$ obtained from the usual theory (i.e., by inverting the information matrix) can be shown to be

$$\text{se}(\hat{\beta}) = \left(\frac{1}{N_s} + \frac{1}{N_i} \right)^{\frac{1}{2}} . \qquad (5.4)$$

This leads to the age estimate

$$\hat{t} = \frac{1}{\lambda} \log \left(1 + \frac{1}{2} \lambda \hat{\zeta} \hat{\rho}_d e^{\hat{\beta}} \right), \qquad (5.5)$$

which is equivalent to (3.12). Also, from (5.1), the standard error of $\hat{\beta}$ approximately equals the *relative* standard error of $\hat{\rho}_s / \hat{\rho}_i$. Hence the relative standard error of the age estimate, including the errors in estimating β, ζ and ρ_d, is

$$\frac{\text{se}(\hat{t})}{\hat{t}} \approx \left\{ \left(\text{se}(\hat{\beta}) \right)^2 + \frac{1}{N_d} + \left(\frac{\text{se}(\hat{\zeta})}{\hat{\zeta}} \right)^2 \right\}^{\frac{1}{2}} , \qquad (5.6)$$

which is equivalent to (3.13). In other words this theory leads to exactly the same method as in Section 3.5.

Now there is another way to obtain a likelihood function for this problem, which uses the following property of Poisson distributions (noted in the Appendix, Section A.3). Suppose that two counts N_1 and N_2 have independent Poisson distributions with means μ_1 and μ_2 and suppose their total is observed, $N_1 + N_2 = m$, say. Then, conditional on this total, N_1 has a binomial distribution with index m and parameter $\theta = \mu_1 / (\mu_1 + \mu_2)$. Intuitively, this is the probability distribution of N_1 amongst pairs N_1, N_2 such that $N_1 + N_2 = m$.

In this approach, the total count for each grain is regarded as being fixed at its observed value. Let $y_u = N_{su}$ and $m_u = N_{su} + N_{iu}$. Then y_u has a binomial distribution with index m_u and parameter θ, where

$$\theta = \frac{\rho_s / \rho_i}{1 + \rho_s / \rho_i} = \frac{e^{\beta}}{1 + e^{\beta}} . \qquad (5.7)$$

The probability of the observed value of y_u is

$$p(y_u | m_u) = \binom{m_u}{y_u} \theta^{y_u} (1 - \theta)^{m_u - y_u} \qquad (5.8)$$

and the log likelihood is therefore

$$L = \sum_{u=1}^{n} \log \left\{ p(y_u | m_u) \right\} + \text{constant} \qquad (5.9)$$

$$= N_s \log \theta + N_i \log (1 - \theta) + \text{constant} \qquad (5.10)$$

$$= N_s\beta - (N_s + N_i) \log (1 + e^\beta) + \text{constant}, \qquad (5.11)$$

where again N_s and N_i are the total numbers of spontaneous and induced tracks summed over all grains. This log likelihood function is simpler than that given by (5.2). It is a function of just one parameter β (or equivalently θ) which is a function of the ratio ρ_s/ρ_i of track densities, and the only data that enter are the total counts N_s and N_i over all grains.

Now, if we start with (5.11) and derive the maximum likelihood estimate of β and its standard error, it transpires (perhaps surprisingly) that this still leads to (5.3) and (5.4). In other words, for the common age model it is immaterial whether we use the log likelihood function (5.2) or (5.11).

In the following sections and in Chapter 6 we will develop some more complicated models to be fitted by maximum likelihood. We will generally base the likelihood function on the (conditional) binomial distributions rather than the (unconditional) Poisson distributions. This probability model does not depend on ρ_{su} or ρ_{iu} separately, nor on the area A_u of crystal surface. By using this conditional distribution we get rid of the nuisance parameters μ_u — in particular, we avoid having to make any assumptions about how uranium concentrations may vary between grains, which is a great simplification. In theory, conditioning on m_u may sometimes result in a small loss of information, but this is greatly outweighed by the gain in simplicity and robustness.

For analysing binomial counts, it is sometimes convenient to calculate the *empirical logits* which, in the present notation, are defined as

$$z_u = \log \left(\frac{N_{su} + \frac{1}{2}}{N_{iu} + \frac{1}{2}} \right) \qquad (5.12)$$

for $u = 1, 2, \ldots, n$. These are estimates of the log odds parameter β from each grain. They have approximate standard errors given by

$$\sigma_u = \sqrt{\frac{1}{N_{su} + \frac{1}{2}} + \frac{1}{N_{iu} + \frac{1}{2}}}. \qquad (5.13)$$

It is customary to add $\frac{1}{2}$ to each count, as in (5.12), to avoid problems with zero or small counts (e.g., Cox, 1970, p. 33). The empirical logits are useful for simple graphical and numerical analyses. Furthermore, when the counts are reasonably large, z_u is approximately normally distributed with mean β and standard deviation σ_u. For some of the models in the chapter and the next, this property leads to simpler estimation formulae.

5.2 Discrete mixture models

A simple extension of the common age model is a mixture of two ages. We imagine the sample grains to be drawn from a population in which a proportion π_1 have age t_1 and π_2 have age t_2, where $\pi_1 + \pi_2 = 1$, $t_1 < t_2$, and it is not known which grains have which age. That is, we imagine that grains from two different common age populations have been mixed together and information about which population each grain came from is lost. This model might be applicable, for example, to apatites from a sandstone derived from contemporaneous volcanism that also contains basement-derived material having older apatites.

We may wish to estimate one or both of t_1 and t_2. We may also be interested in the mixing proportions π_1 and π_2, though often these may be artefacts of the data collection process and therefore may not describe a real population of interest. This remark applies particularly to zircon ages because, even if the field sampling is in some sense representative of the geological environment, it is usually only possible to count fission tracks for grains that have a fairly limited range of uranium concentrations. Nevertheless, the component ages should still be meaningful quantities.

In more complicated situations, this model might be extended to mixtures of three or four different ages. Early applications are discussed by Hurford et al. (1984), Hayashi (1985), Seward and Rhoades (1986), Kowallis et al. (1986), Naeser et al. (1987) and Brandon (1992) and there are many more recent published applications using both apatite and zircon. When a sample has been hot enough to cause track shortening, a yet more complicated model might be needed. In this context, it is worth noting that zircon requires much higher temperatures than apatite for significant annealing to take place, so that discrete mixture models might be useful for fission track dating of zircons in situations when tracks in apatites from the same location are annealed.

To fix notation, again let $N_{su}, N_{iu}, u = 1, 2, \ldots, n$ denote matched pairs of spontaneous and induced track counts for n grains, obtained by the external detector method, and suppose that the grains were irradiated together and that tracks were counted by a single analyst, so that the same estimates of ρ_d and ζ apply to each grain. Suppose that each grain has one of k different ages, $t_1, t_2, \ldots, t_k$, and imagine that each grain is randomly drawn from a large population in which a proportion π_i have age t_i. Then, by the second approach of the previous section, fitting this model is equivalent to fitting a mixture of binomial distributions, where each distribution has a different index $m_u = N_{su} + N_{iu}$ and one of k parameters $\theta_1, \theta_2, \ldots, \theta_k$. The log likelihood function L is given by equation (5.9) but now with

$$p(y_u|m_u) = \binom{m_u}{y_u} \sum_{i=1}^{k} \pi_i \theta_i^{y_u} (1 - \theta_i)^{m_u - y_u} . \qquad (5.14)$$

Extending the previous notation, write

$$\beta_i = \log\left(\frac{\theta_i}{1 - \theta_i}\right) \qquad (5.15)$$

and regard L as a function of the unknown parameters $\beta_1, \beta_2, \ldots, \beta_k$ and mixing proportions $\pi_1, \pi_2, \ldots, \pi_k$. Because the mixing proportions are constrained to add to 1, there are just $2k - 1$ parameters to be estimated. Having obtained estimates of $\beta_1, \beta_2, \ldots, \beta_k$ and their standard errors, the estimates of the ages

$t_1, t_2, \ldots, t_k$ and their relative standard errors are found using equations (5.5) and (5.6).

There is a substantial statistical literature on finite mixture models, dealing with methods of fitting them, statistical properties of the estimates, computational procedures and sources of software. Well-known reference books include Everitt and Hand (1981), Titterington *et al.* (1985) and McLachlan and Peel (2000). These books give general estimation formulae but not explicit formulae for the specific case of interest here, wherein each binomial distribution has a different index. We first look at some examples and then in Section 5.5 we give a simple algorithm for calculating maximum likelihood estimates, along with their precisions and goodness-of-fit diagnostics. In practice it is important to check that the fitted model provides a good description of the observed data.

When the fission track counts are large enough, an approximate log likelihood function can be obtained using z_u and σ_u given by (5.12) and (5.13). Here z_u is from a mixture of Normal (or Gaussian) distributions where each distribution has a different and known standard deviation given by σ_u and one of k unknown means $\beta_1, \beta_2, \ldots, \beta_k$. These are again estimated by maximum likelihood along with the mixing proportions $\pi_1, \pi_2, \ldots, \pi_k$. This can lead to simpler formulae, which we also give in Section 5.5. Sometimes we may wish to fit models directly to the estimated ages and their standard errors (for example if the raw counts are not available). The same approach is applicable here, where z_u is the natural log of the age estimate and σ_u is the relative standard error of the age estimate. In Section 5.8 we discuss briefly how to adapt the formulae to deal with cases where data come from analyses with different ρ_d or ζ values.

5.3 Example: a synthetic mixture of two ages

Table 5.1 shows fission track counts from a laboratory experiment designed to produce a two-age mixture artificially. Apatite grains from the Strontian Granodiorite in Scotland were first heated for one hour at 450°C to remove all spontaneous tracks. Two sub-samples were irradiated to create induced tracks, each with a different neutron fluence, and then grains from each sub-sample were physically mixed together in roughly equal proportions. The two nominal true "ages" in this mixture were approximately 240 Ma and 342 Ma. These were determined independently by measuring the two neutron fluences that were used to generate the induced tracks and converting them to equivalent values in Ma. Forty grains were then "dated" by the external detector method as if they had been obtained in the field. These data are from a series of experiments described in Galbraith and Green (1990). In Table 5.1, and in the other data examples in this chapter, we have not listed the *areas* of grain surface that were counted (cf. Table 3.1) as they do not enter into the likelihood function.

The age estimates are plotted in Figure 5.1, with radial lines corresponding to the two nominal ages 240 and 342 Ma. The data are clearly not consistent

Table 5.1 *Fission track counts obtained by the external detector method for 40 apatite grains from a synthetic mixture, along with single grain ages and standard errors in Ma. Data from Galbraith and Green (1990).*

Grain	N_s	N_i	age (se)	Grain	N_s	N_i	age (se)
1	236	297	321 (28)	21	89	148	244 (33)
2	64	71	362 (62)	22	111	170	265 (32)
3	88	108	328 (47)	23	147	276	217 (22)
4	127	182	282 (33)	24	49	106	188 (33)
5	51	117	178 (30)	25	84	139	245 (34)
6	154	220	283 (30)	26	103	172	243 (30)
7	121	114	425 (55)	27	80	111	291 (43)
8	136	129	422 (52)	28	148	241	249 (26)
9	57	91	254 (43)	29	41	72	231 (45)
10	183	236	313 (31)	30	174	297	238 (23)
11	120	230	212 (24)	31	92	76	482 (75)
12	160	183	352 (38)	32	58	96	245 (41)
13	90	140	261 (35)	33	282	346	329 (26)
14	83	156	216 (29)	34	90	88	410 (61)
15	56	77	294 (52)	35	90	113	321 (45)
16	158	185	344 (37)	36	206	222	373 (36)
17	125	167	302 (36)	37	114	99	459 (63)
18	26	59	180 (42)	38	65	75	349 (59)
19	80	95	339 (51)	39	108	145	301 (38)
20	145	173	338 (38)	40	142	160	357 (41)

N_s = number of spontaneous tracks counted; N_i = number of induced tracks counted; $\hat{\zeta}$ = 353.5 Ma $\times 10^{-6}$ cm^2; relative standard error of $\hat{\zeta}$ = 3.9/353.5 = 0.011; N_d = 5856 tracks; $\hat{\rho}_d$ = 2.34 $\times$ 10^6 cm^{-2}.

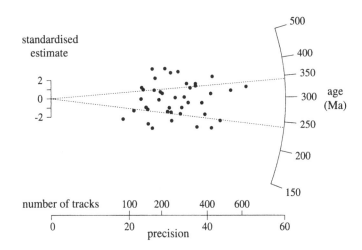

Figure 5.1 *Radial plot of fission track age estimates for the data in Table 5.1. Each grain has one of the two "true" ages 240 Ma or 342 Ma, indicated by the two radii.*

Table 5.2 *Maximum likelihood estimates for a two-age mixture fitted to the data in Table 5.1. The nominal true ages t_i, their estimates $\hat{t}_i$ and their standard errors $se(\hat{t}_i)$ are in Ma.*

	$\hat{\beta}_i$	$se(\hat{\beta}_i)$	$\hat{\pi}_i$	$se(\hat{\pi}_i)$	$\hat{t}_i$	$se(\hat{t}_i)$	t_i
$i = 1$	-0.529	0.041	.44	.10	239.3	10.7	240
$i = 2$	-0.161	0.032	.56	.10	342.9	12.5	342

with a single common age, but do look to be consistent with a mixture of two ages. To see this, observe that the points do not appear scatter homoscedastically with unit standard deviation about a single radius, but practically all points are within ± 2 of one or other of the two radii, with homoscedastic scatter about each.

Maximum likelihood estimates of the parameters of a mixture of $k = 2$ ages are given in Table 5.2. The estimates of $\beta_1, \beta_2, \pi_1, \pi_2$ and their standard errors were calculated by the method in Section 5.5 below, using equations (5.18)–(5.24). Then the resulting estimates of the two ages t_1 and t_2 in Ma were calculated from equations (5.5) and (5.6).

The age estimates agree very closely with the nominal true values. Clearly the method has worked well. Fitting a mixture model with $k = 3$ to these data leads to an increase of only 1.1 in the maximum log likelihood and some evidence that the three ages are not distinct. In fact we know that the two-age model is correct here, but even if we did not, it would be a clear candidate purely on statistical grounds.

5.4 Example: apatite data from the Bengal Fan

The data in Table 5.3, kindly provided by Paul Green, were obtained from an apatite sample from an offshore bore-hole in the Bay of Bengal. Here they serve to illustrate an example with small spontaneous track counts.

In fact eight of the 38 grains have no spontaneous tracks and five have just one, but they do have reasonable numbers of induced tracks. Grain 12 in particular has $N_s = 0$ and $N_i = 105$, which suggests that it may be rather young. If it had a small number of induced tracks also, it need not be young but might simply have a low uranium content. For most of these grains it does not make much sense to quote a point estimate and standard error, so Table 5.3 gives a 95% confidence interval for the age of each grain calculated from equations (3.22) and (3.23).

Figure 5.2 shows a radial plot. Both this and Table 5.3 suggest that the data are not consistent with a single common age. Formally, the chi-square homogeneity test of Section 3.8 gives $\chi^2_{\text{stat}} = 56.6$ with 37 degrees of freedom and a p-value of 0.02. This would normally suggest that some age heterogeneity was present, though because the counts are so small this p-value may not be reliable.

Table 5.3 *Fission track counts obtained by the external detector method for 38 apatite grains from the Bengal Fan (unpublished data from Paul Green), with 95% confidence intervals for single grain ages using (3.22) and (3.23).*

Grain	N_s	N_i	age (Ma)	Grain	N_s	N_i	age (Ma)
1	2	40	1.4 – 47	20	0	8	0.0 –109
2	1	89	0.1 – 16	21	8	135	6.1 – 29
3	1	15	0.4 –104	22	10	218	5.3 – 21
4	24	578	6.4 – 15	23	5	55	6.9 – 55
5	2	36	1.6 – 52	24	0	12	0.0 – 69
6	2	28	2.0 – 68	25	0	127	0.0 – 6
7	14	286	6.4 – 20	26	15	328	6.1 – 19
8	4	93	2.8 – 28	27	5	61	6.2 – 49
9	2	162	0.4 – 11	28	10	341	3.4 – 13
10	16	196	11.1 – 33	29	0	11	0.0 – 76
11	4	114	2.3 – 22	30	11	207	6.3 – 24
12	0	105	0.0 – 7	31	0	39	0.0 – 19
13	3	47	3.1 – 48	32	5	276	1.4 – 10
14	1	74	0.1 – 19	33	5	295	1.3 – 10
15	2	21	2.6 – 94	34	1	57	0.1 – 25
16	0	15	0.0 – 54	35	3	65	2.3 – 34
17	0	19	0.0 – 41	36	2	19	2.9 –105
18	23	373	9.4 – 23	37	18	402	6.4 – 17
19	1	14	0.4 – 13	38	2	115	0.5 – 16

N_s = number of spontaneous tracks counted; N_i = number of induced tracks counted; $\hat{\zeta} = 353.5$ Ma $\times 10^{-6}$ cm^2; relative standard error of $\hat{\zeta} = 3.9/353.5 = 0.011$; $N_d = 2164$ tracks; $\hat{\rho}_d = 1.375 \times 10^6$ cm^{-2}.

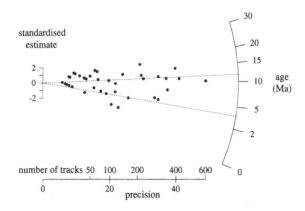

Figure 5.2 *Radial plot of fission track age estimates for the data in Table 5.3.*

Fitting a two-age mixture to these data gives an increase in the maximum log likelihood of 5.12, compared with that for the common age model. The standard likelihood ratio test would compare twice this value, 10.24, with a $\chi^2(2)$ distribution to give a p-value of 0.006, representing strong evidence

Table 5.4 *Maximum likelihood estimates for a two-age mixture fitted to the Bengal Fan data (Table 5.3).*

	$\hat{\beta}_i$	$\text{se}(\hat{\beta}_i)$	$\hat{\pi}_i$	$\text{se}(\hat{\pi}_i)$	$\hat{t}_i$	$\text{se}(\hat{t}_i)$
$i = 1$	-3.864	0.336	.28	.12	3.65	1.23
$i = 2$	-2.188	0.089	.72	.12	11.72	1.04

against the single-age model. Although the usual theoretical conditions for this test to be valid do not strictly apply here (see Section 5.6) this is still clearly suggestive that there is more than one true age present.

Maximum likelihood estimates for the two-age model are given in Table 5.4. The younger age estimate is rather imprecise and the older one has a relative standard error of about 9% but this model is certainly tenable and possibly has a useful interpretation. There is no evidence that more than two distinct ages are present. Indeed, when we try to fit three ages we find evidence of parameter redundancy (i.e., estimates of two of the ages being indistinguishable) and no increase in log likelihood. Also the Bayes Information Criterion (see Section 5.6) is higher for both the single age and the three age models, again supporting the two-age model, at least amongst finite mixtures.

5.5 Maximum likelihood estimation formulae

In this section we use a slightly more general notation so that the formulae may be adapted to mixtures of distributions other than binomial. The subscript u indexes the units (in this case, grains) and i indexes mixture components. The data are independent observations $y_1, y_2, \ldots, y_n$ in which y_u is generated from a distribution with probability mass function (or probability density function in the case of a continuous distribution) of the form

$$f_u(y_u) = \sum_{i=1}^{k} \pi_i f_{iu}(y_u; \beta_i), \qquad (5.16)$$

where k is assumed to be known and $\pi_1 + \pi_2 + \cdots + \pi_k = 1$. Here f_{iu} is a probability mass function or probability density function that depends on an unknown parameter β_i and which may also differ between units. The special cases we consider later are binomial distributions with different indexes (where β_i is the log odds parameter) and Normal distributions with different standard deviations (where β_i is the mean). The log likelihood function is

$$L = \sum_{u=1}^{n} \log f_u(y_u) + \text{constant} \qquad (5.17)$$

considered as a function of $\pi_1, \ldots, \pi_k$ and $\beta_1, \ldots, \beta_k$. Then it can be shown that the maximum likelihood estimates of the βs and πs satisfy the following equations, for $u = 1, 2, \ldots, n$ and $i = 1, 2, \ldots, k$:

$$p_{iu} = \frac{\pi_i f_{iu}}{\sum_{i=1}^{k} \pi_i f_{iu}}, \tag{5.18}$$

$$\pi_i = \frac{1}{n} \sum_{u=1}^{n} p_{iu} \tag{5.19}$$

and

$$\sum_{u=1}^{n} p_{iu} a_{iu} = 0, \tag{5.20}$$

where

$$a_{iu} = \frac{\partial \log f_{iu}}{\partial \beta_i}. \tag{5.21}$$

These equations can be solved numerically by guessing some rough starting values and iterating around (5.18), (5.19) and (5.20). Once the iteration has converged, one may calculate an approximate covariance matrix for the vector of estimates of $(\pi_1, \ldots, \pi_{k-1}, \beta_1, \ldots, \beta_k)$, omitting the redundant parameter $\pi_k = 1 - \pi_1 - \cdots - \pi_{k-1}$. This covariance matrix is given by

$$\begin{pmatrix} \sum_u A_u & \sum_u B_u \\ \sum_u B_u^T & \sum_u C_u \end{pmatrix}^{-1}, \tag{5.22}$$

where A_u is a $(k-1) \times (k-1)$ matrix with (i, j) element

$$\left(\frac{p_{iu}}{\pi_i} - \frac{p_{ku}}{\pi_k} \right) \left(\frac{p_{ju}}{\pi_j} - \frac{p_{ku}}{\pi_k} \right), \quad i, j = 1, 2, \ldots, k-1,$$

B_u is a $(k-1) \times k$ matrix with (i, j) element

$$p_{ju} a_{ju} \left(\frac{p_{iu}}{\pi_i} - \frac{p_{ku}}{\pi_k} - \frac{\delta_{ij}}{\pi_j} + \frac{\delta_{jk}}{\pi_k} \right), \quad i = 1, 2, \ldots, k-1, \; j = 1, 2, \ldots, k$$

and C_u is a $k \times k$ matrix with (i, j) element

$$p_{iu} p_{ju} a_{iu} a_{ju} - \delta_{ij} b_{iu} p_{iu}, \quad i, j = 1, 2, \ldots, k,$$

where $\delta_{ij} = 1$ if $i = j$ and 0 if $i \neq j$, and

$$b_{iu} = \frac{\partial^2 \log f_{iu}}{\partial \beta_i^2} + a_{iu}^2. \tag{5.23}$$

5.5.1 Binomial mixtures

For binomial mixtures, such as those fitted in Sections 5.3 and 5.4, f_{iu} is the binomial probability mass function with index m_u and parameter θ_i and β_i is given by (5.15). So from (5.14), (5.21) and (5.23), f_{iu}, a_{iu} and b_{iu} are given by

$$f_{iu} = \binom{m_u}{y_u} \theta_i^{y_u} (1 - \theta_i)^{m_u - y_u}$$

$$a_{iu} = y_u - \theta_i m_u$$

$$b_{iu} = (y_u - \theta_i m_u)^2 - \theta_i (1 - \theta_i) m_u.$$

Then equation (5.18) becomes

$$\theta_i = \frac{\sum_{u=1}^n p_{iu} y_u}{\sum_{u=1}^n p_{iu} m_u}. \tag{5.24}$$

A simple iterative scheme for calculating the estimates is:

1. Assign starting values to π_i and θ_i, $i = 1, \ldots, k$.
2. Calculate p_{iu}, $i = 1, \ldots, k$, $u = 1, \ldots, n$ using (5.18).
3. Update π_i, $i = 1, \ldots, k$ using (5.19).
4. Update θ_i, $i = 1, \ldots, k$ using (5.24).
5. Calculate L from (5.17).

Then repeat steps 2–5 until the estimates stabilise and L is maximised. The final values of θ_i, $i = 1, \ldots, k$ can then be used to calculate estimates of β_i, $i = 1, \ldots, k$ from (5.15). Then the approximate covariance matrix, and hence standard errors, may be calculated from (5.22). Section A.10 in the Appendix indicates how to obtain standard errors of maximum likelihood estimates.

This algorithm should produce estimates that maximise L given by (5.17) but it is wise to compute L for a range of other parameter values as a check and also to try different starting values in step 1 to ensure that the same solution is reached. Experience suggests that the choice of starting values is not critical, provided that the number of ages being fitted is not too large. Simple default starting values are to set the mixing proportions all equal (i.e., $\pi_i = 1/k$) and the log odds parameters β_i equally spaced over the range of empirical logits, e.g.,

$$\beta_i = z_{(1)} + \tfrac{i}{k+1}(z_{(n)} - z_{(1)}),$$

where $z_{(1)}$ and $z_{(n)}$ are the smallest and largest z_u values in (5.12). Corresponding values of θ_i may be obtained from (5.7). When calculating f_{iu} and L there is no need to calculate the binomial coefficients or their logarithms because they do not depend on the unknown parameters. Also it helps to add a suitable constant to L to get a value of a convenient size.

It can sometimes be useful to print a table of the final values of p_{iu} to 2 decimal places. These indicate which mixture component or components each grain is likely to come from.

5.5.2 Normal mixtures

Suppose that we have independent observations $z_1, z_2, \ldots, z_n$ where z_u is from a Normal distribution with known standard deviation σ_u (which may differ for each unit or grain) and with unknown mean taking the value β_i with probability π_i for $i = 1, 2, \ldots, k$. This is analogous to the binomial mixture considered above and, for suitably defined z_u, can sometimes provide an alternative simpler method of estimation. Writing $x_u = 1/\sigma_u$ and $y_u = z_u/\sigma_u$, this model is equivalent to y_u being from a mixture of Normal distributions with probability density function given by (5.16) where f_{iu} is the Normal probability density function with mean $\beta_i x_u$ and unit standard deviation. Then f_{iu}, a_{iu} and b_{iu} are given by

$$f_{iu} = (2\pi)^{-\frac{1}{2}} \exp\left(-\tfrac{1}{2}(y_u - \beta_i x_u)^2\right)$$

$$a_{iu} = x_u(y_u - \beta_i x_u)$$

$$b_{iu} = -x_u^2\left(1 - (y_u - \beta_i x_u)^2\right).$$

Equations (5.18), (5.19) and (5.20) still apply, where (5.20) becomes

$$\beta_i = \frac{\sum_{u=1}^n p_{iu} x_u y_u}{\sum_{u=1}^n p_{iu} x_u^2}. \tag{5.25}$$

The maximum likelihood estimates and their standard errors can be calculated in the same way as before:

1. Assign starting values to π_i and β_i, $i = 1, \ldots, k$.
2. Calculate p_{iu}, $i = 1, \ldots, k$, $u = 1, \ldots, n$ using (5.18).
3. Update π_i, $i = 1, \ldots, k$ using (5.19).
4. Update β_i, $i = 1, \ldots, k$ using (5.25).
5. Calculate L from (5.17).

Then repeat steps 2–5 until the estimates stabilise and L is maximised. Again, the approximate covariance matrix, and hence standard errors, may be calculated from (5.22) using the final parameter values.

If the spontaneous and induced track counts N_{su} and N_{iu} are reasonably large, then these formulae may be applied with z_u and σ_u given by (5.12) and (5.13). This provides an alternative method for fitting a mixture of k ages. Writing $x_u = 1/\sigma_u$ and $y_u = z_u/\sigma_u$ we may apply equations (5.18), (5.19) and (5.25). The parameters β_i have exactly the same interpretation as before and age estimates may again be calculated using (5.5) and (5.6).

Applying this method to the data in Table 5.1 gives the estimates in Table 5.5 for a two-age mixture. These are practically identical to the estimates in Table 5.2 from the binomial mixture model. But for the data in Table 5.3,

Table 5.5 *Maximum likelihood estimates using the Normal model for the data in Table 5.1 (c.f. Table 5.2).*

	$\hat{\beta}_i$	se($\hat{\beta}_i$)	$\hat{\pi}_i$	se($\hat{\pi}_i$)	$\hat{t}_i$	se($\hat{t}_i$)
$i = 1$	−0.526	0.042	.44	.10	239.8	10.8
$i = 2$	−0.161	0.032	.56	.10	342.9	12.5

where the counts are small, this Normal approximation breaks down and does not give sensible estimates.

This Normal mixture model is also useful more generally, for example when the available data are age estimates themselves and their standard errors, rather than the raw counts. Here we may define z_u to be the natural log of the age estimate and σ_u to be the *relative* standard error of the age estimate, with $x_u = 1/\sigma_u$ and $y_u = z_u/\sigma_u$. This method is illustrated in Section 5.7.

5.6 How many ages to fit?

A common problem in applications is to decide how many distinct age components are present. The importance of this question will depend on the context. Sometimes it may be of prime interest to discover whether there are, say, three rather than two distinct ages, perhaps indicative of different geological events. Other times the main interest may focus on estimating a particular age component, perhaps the youngest, and if essentially the same estimate of the youngest age is obtained by fitting three and by fitting four ages, the question of how many distinct ages are really present may be of little consequence.

Sometimes the geological context may help to determine k, or at least a minimum value for k. In the absence of hard information, a natural strategy is to fit successively $k = 1, 2, 3, \ldots$ ages until a good fit is obtained and no material improvement results from fitting more. More often than not in practice this will lead to a sensible choice with little ambiguity. In particular, attempting to fit more age components than are supported by the data often results in evidence of either parameter redundancy or numerical problems (or both). Parameter redundancy can be indicated by two of the age estimates being identical or very close, or one of the mixing proportions being close to zero. Numerical problems include lack of convergence in the estimation algorithm to a single solution or singularity in the calculated information matrix. Either of these can suggest that there are too many components in the model being fitted.

There is a considerable, and still developing, literature on formal statistical methods for estimating k in a finite mixture. McLachlan and Peel (2000, §5.11) give a recent brief review. One criterion used for this purpose is the increase in $2L_{max}$ achieved by fitting $k+1$ ages rather than k, where L_{max} is the maximum value of the log likelihood. If after fitting k components a further component is fitted, $2L_{max}$ must increase in principle; but if the extra component is not supported by the data the increase will be small. A standard log likelihood ratio test would assess the significance of the increase by comparing it with the $\chi^2(2)$ distribution (because fitting $k+1$ rather than k ages adds 2 parameters to the model). Strictly speaking, the mathematical conditions do not hold for the standard theory to apply without modification, as a model with fewer ages may be obtained by putting one or more mixing proportions equal to zero, which is at the boundary of possible values. Nevertheless, this test is still a useful guide.

Another useful criterion is BIC, the Bayes Information Criterion of Schwartz (1978). This is defined as

$$BIC = -2L_{max} + p \log n, \qquad (5.26)$$

where L_{max} is the maximised value of the log likelihood, p is the number of fitted parameters, and n is the number of grains. Here $p = 2k - 1$ for a k-age model. If a mixture model with more ages is fitted, L_{max} must increase (or at least not decrease) so the first term on the right-hand side of (5.26) will decrease and the second term will increase. As k increases, BIC will initially decrease to a point where the increase in log likelihood is outweighed by the increase in number of parameters fitted, after which BIC will increase again. A simple rule is therefore to choose the k that gives the smallest BIC.

Experience suggests that it is usually quite easy to find the appropriate number of ages using these criteria. In practice, though, it is wise not to rely solely on formal statistical criteria such as L_{max} or BIC, but rather to take the context and purpose into account. One must always be aware that several different models may provide an equally good fit, in statistical terms, to the same data.

5.7 Example: zircon ages from Mount Tom

The single grain age estimates in Table 5.6 illustrate an example with more than two component ages. These age estimates for 50 zircon grains, along with *twice* their standard errors, were presented by Brandon (1992). The radial plot of these data (Figure 5.3) indicates visually that they would be well explained by four distinct ages.

Here we do not have the original track counts available, so we will fit Normal mixture models by the method in Section 5.5.2 with z_u equal to the log of the observed age and σ_u equal to the relative standard error, se(age)/age, for grain u. The component age estimates are then given directly as $\hat{t}_i = e^{\hat{\beta}_i}$ and their approximate relative standard errors are $se(\hat{t}_i)/\hat{t}_i = se(\hat{\beta}_i)$.

Table 5.6 *Observed single grain age (Ma) and 2 × standard error (Ma) for 50 zircon grains from Mount Tom, Washington, USA. Data from Brandon (1992, Table 1).*

age	2 se	age	2 se	age	2 se	age	2 se	age	2 se
14.0	4.6	20.6	6.5	41.0	12.9	50.6	15.0	66.7	18.2
14.1	4.1	20.9	7.2	41.5	11.8	54.6	18.7	72.0	22.4
14.1	7.3	21.3	6.3	42.8	12.0	55.2	21.9	73.0	22.6
17.4	6.3	28.2	9.7	44.5	12.4	58.7	17.7	77.8	29.6
18.0	6.0	31.8	16.3	44.8	15.0	58.8	14.1	80.2	24.5
18.9	6.8	35.8	12.0	45.7	13.5	59.8	23.7	81.2	28.3
19.2	8.3	38.7	12.6	45.8	11.7	60.7	16.1	81.4	29.0
20.0	6.2	38.9	9.6	46.3	14.3	63.0	20.9	91.8	30.8
20.0	6.9	40.3	14.1	49.1	10.5	63.6	17.5	153.1	55.4
20.1	9.2	40.8	10.9	50.5	18.6	65.3	18.2	195.0	62.9

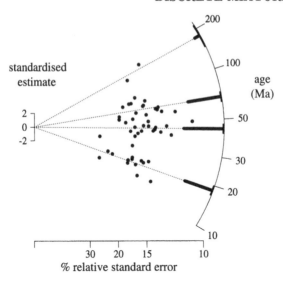

Figure 5.3 *Radial plot of the fission track age estimates (log scale) in Table 5.6. Radial lines indicate estimates of four distinct ages of 19, 44, 67 and 175 Ma. Bars indicate the mixing proportions and 95% confidence intervals for the component ages (Table 5.7).*

Successively fitting $k = 2, 3, 4$, and 5 ages gives the following values of the maximum log likelihood $L_{\max}$ and the Bayes Information Criterion BIC:

k	p	$L_{\max}$	BIC
2	3	23.31	65.10
3	5	62.21	−4.86
4	7	77.19	−27.00
5	9	77.34	−19.47

where BIC is given by equation (5.26) with $p = 2k - 1$ and $n = 50$. The increase in $L_{\max}$ is substantial when going from two to three ages, and from three to four, but is insignificant when going from four to five. Also BIC takes its smallest value when $k = 4$, all of which suggests that four ages are both necessary and sufficient. Age estimates for the four component mixture are

Table 5.7 *Estimates for a four-age mixture fitted to the data in Table 5.6.*

component	1	2	3	4
age (Ma)	18.6	43.9	67.2	175.2
s.e. (Ma)	0.6	1.3	2.1	17.4
proportion	.273	.372	.315	.040
s.e.	.064	.077	.074	.028

given in Table 5.7. These ages are drawn as radii on Figure 5.3, which also confirms that the four-age model fits well.

5.8 Data from more than one irradiation

Sometimes one may wish to fit mixture models to counts obtained from more than one irradiation run. Then ρ_d is not the same for all grains and the above methods may need to be modified appropriately. We mention some simple approximate methods later, but for completeness we first modify the method in Section 5.5.

The data are now of the following form: for each of n_d irradiation runs, $j = 1, 2, \ldots, n_d$, we have a dosimeter count N_{dj} over an area A_{dj} from a dosimeter glass; and n_j pairs of fission track counts N_{sju}, N_{iju} for grains $u = 1, 2, \ldots, n_j$. We wish to fit a mixture of k ages to all grains together.

For grains from a single component i, the counts have independent Poisson distributions with means of the form

$$E(N_{dj}) = A_{dj}\rho_{dj}, \quad E(N_{iju}) = \rho_{dj}\mu_{ju} \quad \text{and} \quad E(N_{sju}) = e^{\beta_i}\mu_{ju},$$

where the μ_{ju} are unknown nuisance parameters. In addition, each grain has probability π_i of being from component i (that is, of having age t_i). Extending the analysis in Section 5.5, let $y_{ju} = N_{sju}$ and $m_{ju} = N_{sju} + N_{iju}$. Then if grain (j, u) is from component i, the conditional probability distribution of y_{ju} given m_{ju} is binomial with index m_{ju} and parameter θ_{ij} where

$$\theta_{ij} = \frac{e^{\beta_i}}{e^{\beta_i} + \rho_{dj}}. \tag{5.27}$$

Now θ_{ij} depends on both β_i and ρ_{dj}.

To fit a discrete mixture of such distributions, let us first agree to estimate the dosimeter track densities ρ_{dj} by the usual formula $\hat{\rho}_{dj} = N_{dj}/A_{dj}$, and then estimate θ_{ij} as if these were known. Then we may use the method based on equations (5.18)–(5.24) where the summations over grains are now summations over both u and j. This is not quite the same as using full maximum likelihood estimation from all of the data because, strictly speaking, in addition to the dosimeter counts, the induced fission track counts also contain some information about how ρ_d varies between irradiations. Formally, using the full likelihood function from both dosimeter and fission track counts would give an estimate of ρ_{dj} that differs from the usual one. But it is desirable to use the standard estimate of ρ_{dj}. Then

$$f_{iju} = \binom{m_{ju}}{y_{ju}} \theta_{ij}^{y_{ju}} (1 - \theta_{ij})^{m_{ju} - y_{ju}}$$

$$a_{iu} = y_{ju} - \theta_{ij} m_{ju}$$

$$b_{iu} = (y_{ju} - \theta_{ij} m_{ju})^2 - \theta_{ij}(1 - \theta_{ij}) m_{ju}.$$

Equations (5.18) and (5.19) now become

$$p_{iju} = \frac{\pi_i f_{iju}}{\sum_{i=1}^{k} \pi_i f_{iju}} \tag{5.28}$$

and

$$\pi_i = \frac{1}{n} \sum_{j=1}^{n_d} \sum_{u=1}^{n_j} p_{iju}, \tag{5.29}$$

where $n = n_1 + n_2 + \cdots + n_d$ is the total number of grains in all irradiations. And in place of (5.24) we have

$$\sum_{j=1}^{n_d} \sum_{u=1}^{n_j} p_{iju} y_{ju} = \sum_{j=1}^{n_d} \sum_{u=1}^{n_j} \theta_{ij} p_{iju} m_{ju}, \tag{5.30}$$

where θ_{ij} is given by (5.27). This last equation does not now give β_i explicitly, but it can be solved numerically at each iteration in the algorithm to update to a new β_i. Having found the final values of the βs and πs, standard errors may be obtained using (5.22) as before.

The above method is quite complicated and it is therefore worth mentioning two simpler approximate methods. First, if the fission track counts are large enough, then one may use the Normal mixture method as above but with z_u now defined by $z_u = \log(\hat{\rho}_d N_{su}/N_{iu})$ and $\sigma_u = \sqrt{1/N_d + 1/N_{su} + 1/N_{iu}}$. Alternatively, define z_u to be the log of the age estimate and σ_u to be its relative standard error, as in Section 5.7. Strictly speaking, counts from grains with the same $\hat{\rho}_d$ will be correlated, but if N_d is large compared to the other counts, which is the usual situation, the effect of this will be too small to matter.

Secondly, if there is only a small number of different irradiation runs each with a reasonably large number of grains, then the component ages may be estimated separately for each run, by whatever method is appropriate. Then an overall estimate of each component age may be obtained as a weighted average of the separate estimates, with weights equal to the reciprocals of their variances. For this method one would want to be sure that the same component ages were present in each sub-sample.

5.9 Bibliographic notes

The formulae in Section 5.5 are based on Galbraith (1988, Appendix 1) and Galbraith and Green (1990).

CHAPTER 6

Continuous mixtures of ages

The type of mixture model discussed in Chapter 5 may not always be appropriate. For example, because track density depends on the mean track length — see equation (3.2) — a spread in fission track ages is often caused by thermal annealing when there is compositional variation. Consider a set of apatite grains undergoing a thermal history such as (b) in Chapter 1, Figure 1.3. Within each grain, tracks formed after t_0 will be shortened by heat, those formed soonest being shortened most. The mean length, and therefore the track density ρ_s, will be less than for thermal history (a), and the age t, as given by the fission track age equation (3.6), will be less than t_0. Nevertheless, if all grains are equally sensitive to heat, ρ_s/ρ_i will be the same for all grains so they will have a common t, albeit less than t_0. Although it is conventional to express results in terms of ages, we are really talking about variation in ρ_s/ρ_i for grains receiving the same neutron dose.

However, if different grains are *not* equally sensitive to heat then the mean length μ_s, and hence ρ_s/ρ_i, will vary between grains. For example it is known that tracks in a pure fluorapatite anneal more quickly than tracks in an apatite containing some chlorine (Green *et al*, 1986). If the amount of chlorine varies randomly between apatite grains, then ρ_s/ρ_i will also vary randomly between grains. In this chapter we consider *random effects* models for this situation. In the simplest model, there is a *distribution* of values of $\log(\rho_s/\rho_i)$ summarised by two parameters: a mean and a standard deviation. These can be interpreted in terms of a central age and an age dispersion.

Similar arguments apply to other thermal histories such as (c) and (d) in Figure 1.3. If the chlorine content varies between apatite grains then so will ρ_s/ρ_i. Note that for thermal history (a), though, there would be practically no variation in ρ_s/ρ_i even if the chlorine contents did vary because the maximum temperatures experienced by practically all tracks are too low to anneal them to any significant degree.

6.1 Example: Otway data from Australia

Figure 6.1 displays fission track age estimates for two samples of apatite from the Otway region in Victoria, Australia. There are 20 grains from an outcrop and 30 grains from a depth of 2.6 km in a bore-hole nearby. Table 6.1 lists the raw counts.

The outcrop estimates are clearly consistent with each other and they agree with the depositional age of about 120 Ma cited by Gleadow and Duddy

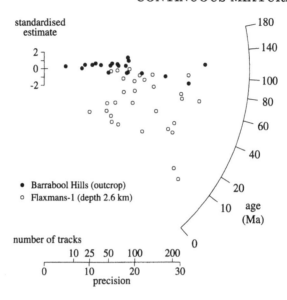

Figure 6.1 *Radial plot of fission track age estimates for two apatite samples from Victoria, Australia. Closed circles are 20 grains from an outcrop (Barrabool Hills, Otway Ranges) and open circles are 30 grains from a depth of 2.6 km (Flaxmans-1 bore-hole, Otway Basin). Data from Gleadow and Duddy (1981).*

Table 6.1 *Spontaneous and induced fission track counts obtained by the external detector method for the data plotted in Figure 6.1.*

Outcrop ($\hat{\zeta}\hat{\rho}_d = 429.7\,\text{Ma}$)						Bore-hole ($\hat{\zeta}\hat{\rho}_d = 369.8\,\text{Ma}$)							
N_s	N_i	N_s	N_i	N_s	N_i	N_s	N_i	N_s	N_i	N_s	N_i	N_s	N_i
28	57	22	34	16	34	36	76	7	186	31	51	22	48
76	177	36	49	8	12	39	94	3	45	8	47	56	154
30	51	25	40	2	3	46	85	2	47	35	67	18	32
115	199	13	19	35	52	16	54	0	22	23	68	3	56
6	11	20	33			59	149	3	39	9	95	25	154
11	17	27	56			22	38	28	63	24	143	26	111
25	42	15	24			83	150	9	172	5	38		
57	121	38	78			72	194	32	153	16	70		

(1981). This homogeneity is typical of samples with a thermal history such as in Figure 1.3 (a), i.e., rapid cooling from a high temperature to below about 50°C and no subsequent heating. So here the fission track age indicates the time of cooling.

The bore-hole estimates differ in two respects: they are generally younger and they are heterogeneous, with fission track ages ranging from zero to 120 Ma. These are consequences not of later deposition, but rather of thermal annealing due to the higher temperature experienced at this depth (currently about 92°C). Geological constraints imply that the depositional age of this sample cannot be younger than the stratigraphic age of the outcrop. So here the fission track ages are telling us indirectly something about the amount of thermal annealing the sample has undergone. The heterogeneity is due to varying chemical composition between crystals, those with more chlorine being less sensitive to heat. In this situation, geological interpretation is greatly clarified by measuring track *lengths* in addition to densities, which we discuss in later chapters. Here we are trying to summarise and interpret the fission track ages.

6.2 General approach

Starting with equation (3.5) and substituting $\sigma_f I \Phi / \lambda_f = \zeta \rho_d$ (i.e., using the zeta calibration method) leads to

$$t = \frac{1}{\lambda} \log \left(1 + \frac{1}{2} \lambda \zeta \rho_d \frac{\rho_s}{\rho_i} \frac{\mu_i}{\mu_s} \right). \tag{6.1}$$

This differs from the usual fission track age equation (3.10) in that the term μ_i / μ_s is still included. This is the ratio of the equivalent isotropic lengths of induced and spontaneous tracks, which in annealed samples may differ from unity — and indeed may vary between grains. As in Chapter 5, define

$$\beta = \log \left(\frac{\rho_s}{\rho_i} \right). \tag{6.2}$$

Then equation (6.1) may be rearranged as follows:

$$\beta = \log 2 - \log \zeta - \log \rho_d + \log \left(\frac{\mu_s}{\mu_i} \right) + \log \left(\frac{e^{\lambda t} - 1}{\lambda} \right). \tag{6.3}$$

This shows how the true value of β for a single grain depends on ζ, ρ_d, μ_s / μ_i and t. The last term on the right-hand side is approximately $\log t$.

Usually ζ and ρ_d are the same for all grains in a sample or suite of samples. But μ_s / μ_i may vary between grains if tracks have partially annealed by differing amounts; and t will vary between grains if tracks have been forming over different time periods. More generally, when combining data from different irradiations ρ_d may vary; or when combining data from different observers ζ may also vary between grains. It is natural to combine all of these sources of variation additively on the log scale. Furthermore variations in ρ_s and ρ_i are then treated symmetrically. Also ρ_s / ρ_i is necessarily positive, but β is unrestricted and can be simply modelled, for example by a Normal distribution, with parameters that have a clear meaning. We therefore specify models for how β varies and convert the results to the age scale for interpretation.

6.3 A random effects model with binomial errors

As in Chapter 5, suppose the data are spontaneous and induced fission track counts N_{su}, N_{iu}, $u = 1, 2, \ldots, n$ for n grains, obtained by the external detector method in a single irradiation and counted by a single analyst. Write $y_u = N_{su}$ and $m_u = N_{su} + N_{iu}$ and let

$$\beta_u = \log(\rho_{su}/\rho_{iu}),\qquad\qquad(6.4)$$

where ρ_{su} and ρ_{iu} are the true spontaneous and induced track densities for grain u. In the random effects model it is assumed that β_u is randomly drawn from a Normal population with mean μ and standard deviation σ and then the spontaneous track count y_u is drawn from a binomial distribution with index m_u and parameter $\theta_u = e^{\beta_u}/(1 + e^{\beta_u})$. The count y_u is observed, but the value of β_u is not.

In physical terms, we may imagine the sample grains to be randomly drawn from a population of grains, each of which has an associated notional value of ρ_s/ρ_i and hence of $\beta = \log(\rho_s/\rho_i)$. The assumption that β is Normal is of course arbitrary but more justifiable than assuming, for example, that ρ_s/ρ_i is Normal, for the reasons given above. The parameters σ and μ correspond to the standard deviation and mean of β for the whole population of grains. The parameter σ is usually called the *age dispersion* and the fission track age that corresponds to μ is called the *central age* — i.e., the age corresponding to $\rho_s/\rho_i = e^\mu$ in equation (3.10), which we denote by t_μ.

Because ρ_s/ρ_i is, to a close approximation, proportional to t, the model implies that the population distribution of true single grain ages is approximately *log-normal*, and therefore positively skewed. The age dispersion is effectively the *relative* standard deviation (or coefficient of variation) of this distribution and the central age is effectively its geometric mean. In an example such as the bore-hole data in Figure 6.1, the central age does not correspond to the time of cooling of the sample. Nor do the single grain ages of course. Rather, the central age and dispersion together summarise the variation in true fission track ages t defined by equation (3.10). In applications, the age dispersion is often quoted as a percentage: for example, $\sigma = 0.15$ means that the true fission track ages of single grains vary with a standard deviation of 15% of their central age.

This model was introduced in Chapter 3, see Section 3.9, where we gave a simple algorithm for estimating μ and σ. When $\sigma = 0$ the random effects model reduces to the common age model and the central age estimate then coincides with the pooled age (see Figure 3.5). More generally when σ is small the central age estimate is very close to the pooled age, but has a slightly larger standard error. In general when estimating σ and μ there are two sources of error. Firstly we only have a *sample* of grains from the population, so even if we could measure β exactly for each grain in the sample we would not know σ and μ exactly. Secondly, because of the Poisson errors associated with the track counts we can only *estimate* β for each grain. From about 20 grains one can usually get a fairly precise estimate of μ, and hence of t_μ, but rather more grains than this are typically needed to obtain a precise estimate of σ. The precisions will also depend on the numbers of tracks counted per grain, of course. Below we give maximum likelihood estimation formulae and look at some examples.

6.4 Maximum likelihood estimation formulae

The random effects model specifies a probability for any observed sample of counts, in terms of the unknown parameters σ and μ. This probability is the product of the independent probabilities for each grain. For grain u the probability is $p(y_u|m_u)$ given by

$$p(y_u|m_u) = \binom{m_u}{y_u} \int_{-\infty}^{\infty} e^{\beta y_u}[1 + e^{\beta}]^{-m_u} \frac{1}{\sigma\sqrt{2\pi}} e^{-\frac{1}{2}(\frac{\beta-\mu}{\sigma})^2} \, d\beta. \qquad (6.5)$$

The log likelihood L, regarded as a function of the two parameters μ and σ, is then

$$L = \sum_{u=1}^{n} \log \left\{ p(y_u|m_u) \right\} + \text{constant},$$

with $p(y_u|m_u)$ given by (6.5). The integral in (6.5) cannot be evaluated explicitly, so it needs to be calculated numerically for given values of μ and σ. Hence, calculation of L typically requires a separate numerical integration for each grain for each μ and σ.

This statistical model has been used in other contexts and some numerical and approximate methods are given in the literature — e.g., Bock and Lieberman (1970), Williams (1982), Anderson (1988) and Goutis (1993). Maximum likelihood estimates, denoted by $\hat{\sigma}$ and $\hat{\mu}$, are those values of μ and σ for which L is largest. These can be computed either by using a numerical optimisation program or by computing L for a grid of values of μ and σ and locating the maximum by inspection. To aid convergence and to obtain a more reliable standard error, it is generally better to work with $\kappa = \log \sigma$, maximise L over μ and κ, and hence obtain $\hat{\sigma} = e^{\hat{\kappa}}$.

Approximate standard errors for $\hat{\mu}$ and $\hat{\kappa}$, and hence for $\hat{\sigma}$, can be obtained by calculating the 2×2 matrix

$$\begin{pmatrix} v_{11} & v_{12} \\ v_{12} & v_{22} \end{pmatrix} = \begin{pmatrix} -\frac{\partial^2 L}{\partial \mu^2} & -\frac{\partial^2 L}{\partial \mu \partial \kappa} \\ -\frac{\partial^2 L}{\partial \mu \partial \kappa} & -\frac{\partial^2 L}{\partial \kappa^2} \end{pmatrix}^{-1},$$

where the second derivatives of the log likelihood L also need to be calculated numerically for $\mu = \hat{\mu}$ and $\kappa = \hat{\kappa}$. Then the approximate standard errors are

$$\text{se}(\hat{\mu}) \approx \sqrt{v_{11}} \quad \text{and} \quad \text{se}(\hat{\kappa}) \approx \text{se}(\hat{\sigma})/\sigma \approx \sqrt{v_{22}}.$$

An alternative, and perhaps more reliable, method for inferring the precision of the estimate of σ is to plot its profile log likelihood function (see Appendix Section A.10).

6.5 A random effects model with Normal errors

When the numbers of tracks per grain are reasonably large, more explicit formulae that are easier to compute can be obtained using the empirical logits, i.e., z_u and σ_u given by (5.12) and (5.13). Here we assume that z_u is drawn from a Normal population with mean β_u and standard deviation σ_u, where, as before, β_u is randomly drawn from a Normal population with mean μ and standard deviation σ. Then together these imply that z_u is drawn from a Normal population with mean μ and standard deviation $\sqrt{\sigma^2 + \sigma_u^2}$. The probability density of z_u is thus

$$f_u(z_u) = \frac{1}{\sqrt{2\pi(\sigma^2 + \sigma_u^2)}} \exp\left(-\frac{(z_u - \mu)^2}{2(\sigma^2 + \sigma_u^2)}\right). \tag{6.6}$$

Then the log likelihood function is

$$L = \sum_{u=1}^{n} \log f_u(z_u) + \text{constant}$$

$$= -\frac{1}{2}\sum_{u=1}^{n} \left\{\log(\sigma^2 + \sigma_u^2) + \frac{(z_u - \mu)^2}{\sigma^2 + \sigma_u^2}\right\} + \text{constant} \tag{6.7}$$

after rearranging. As a function of μ and σ, this is rather more tractable than the log likelihood based on (6.5) and no numerical integrations are needed to compute it.

From (6.7) it can be shown that the maximum likelihood estimates of μ and σ satisfy the equations

$$\mu = \frac{\sum_{u=1}^{n} w_u z_u}{\sum_{u=1}^{n} w_u} \tag{6.8}$$

and

$$\sum_{u=1}^{n} w_u^2 (z_u - \mu)^2 = \sum_{u=1}^{n} w_u, \tag{6.9}$$

where $w_u = 1/(\sigma^2 + \sigma_u^2)$. Equations (6.8) and (6.9) can be solved numerically to obtain the estimates of μ and σ, denoted by $\hat{\mu}$ and $\hat{\sigma}$. These estimates have approximate standard errors given by

$$\text{se}(\hat{\mu}) = \left(\sum_{u=1}^{n} w_u\right)^{-\frac{1}{2}} \quad \text{and} \quad \text{se}(\hat{\sigma}) = \left(2\sigma^2 \sum_{u=1}^{n} w_u^2\right)^{-\frac{1}{2}},$$

which can be estimated by substituting $\hat{\sigma}$ in place of σ. Therefore, for large fission track counts, approximate maximum likelihood estimates of the central age and age dispersion, and their standard errors, are easy to calculate.

Furthermore it is easy to compute the profile log likelihood for σ because, once σ is fixed, $\hat{\mu}$ is determined explicitly. Thus, for each potential value of σ, compute $w_u = 1/(\sigma^2 + \sigma_u^2)$, then μ from equation (6.8) and then L from (6.7). The constant in (6.7) can be chosen so that L is a convenient size. An approximate 95% confidence interval for σ then consists of those values for which L is within 1.92 of its maximum value (see Appendix, Section A.10).

Although we have introduced this Normal model as an approximation to the binomial model of Section 6.4 it is of course a useful model in its own right, applicable when the estimation errors of the single grain ages (or some function of them) can be assumed to be normally distributed. For example, it might be applied directly to the log age estimates, for data such as those in Section 5.7.

6.6 Examples

Let us look again at the Otway data in Table 6.1 and Figure 6.1. For the outcrop sample, where there is no evidence of heterogeneity, the estimate of σ converges to 0 and the central age estimate coincides with the pooled age of 113.0 Ma with standard error 5.8 Ma. But for the bore-hole data, where the single grain ages are highly heterogeneous, the maximum likelihood estimate of σ is 0.80 and the corresponding central age estimate is 41.1 Ma. In other words, it is estimated that the *true* single grain fission track ages vary about 41.1 Ma with a coefficient of variation of about 80%. Remember that because of the Poisson variation in track counts, the *observed* fission track ages will vary by more than this.

Approximate 95% confidence intervals for σ and t_μ, calculated from their profile log likelihood functions, are 0.59–1.12 and 29.8–56.6 Ma, respectively. These indicate how precisely the two parameters are estimated. For interest, the corresponding calculations using the Normal model in Section 6.5 give an age dispersion of 0.71 (95% confidence interval 0.51–0.99) and a central age of 45.3 Ma (with relative standard error 6.44% and a 95% confidence interval 39.9–51.6 Ma). The two methods do differ a little, no doubt because some of the counts in Table 6.1 are small, which would make the Normal method less reliable.

To illustrate how the Normal model may apply when the counts are reasonably large, let us use the data in Table 5.1. Of course we know that the random effects model is not right for these data which are a mixture of two ages, but from a methodological viewpoint it is interesting to see if the same estimates are obtained using (6.5) and (6.6). For these data, the maximum likelihood estimates of the central age and dispersion, based on (6.5) are $\hat{t}_\mu = 294.4$ Ma and $\hat{\sigma} = 0.198$, while those based on the Normal model (6.6) are $\hat{t}_\mu = 294.9$ Ma and $\hat{\sigma} = 0.197$, which are practically identical. For counts as large as those in Table 5.1 the Normal model works very well.

A good example of the use of the random effects model is shown in Figure 6.2. This displays age estimates for samples of apatite crystals from four depths in a well along with summary statistics consisting of the sample depth, the present-day temperature, the central age and the age dispersion (quoted as the standard deviation of the true single grain ages as a percentage of their central age). The pattern is striking. The estimated ages decrease as depth

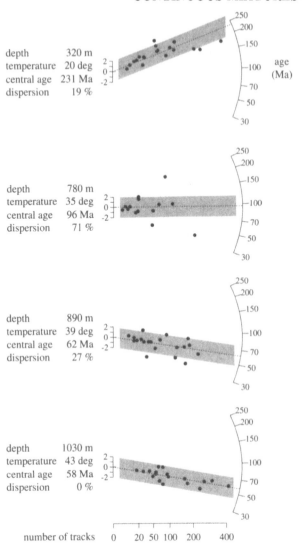

Figure 6.2 *Radial plots of fission track age estimates and summary statistics for apatite samples from four depths in a well. The radial lines show the central age for each sample. Data from Green (1989).*

increases, as shown by the lines corresponding to central ages. Also in the top three samples, the estimates are over-dispersed, with several points outside the ±2 shaded bands, suggesting that there is some variation in true fission track ages. The dispersion increases and then decreases as the central age decreases. In the bottom sample the age dispersion is zero and the scatter is consistent with a common fission track age for all grains.

This pattern can be explained by the combined effects of increasing temperature with depth and varying chemical composition of the apatite grains. As the temperature increases and tracks become shorter, ρ_s/ρ_i will both decrease and become more variable, to a point where tracks in some crystals (those with more chlorine) are only slightly shortened while tracks in others are considerably shortened. At higher temperatures tracks become still shorter so that all grains have mostly short tracks, then ρ_s/ρ_i will decrease further but will be less variable.

The current temperatures for the four samples are shown in Figure 6.2. They increase with depth but are too low to produce significant annealing of tracks. One may therefore infer that the temperatures have been higher in the past, as well as increasing with depth. Such effects are usually due to deeper burial followed by exhumation, but may also be caused by hot fluids, elevated heat flow or other thermal events. In the bottom sample the ages are homogeneous and consistent with a common age of 58 Ma. The inference here is that this sample has been hot enough to obliterate practically all tracks formed prior to 58 Ma and since then has been cool enough that little or no fission track age reduction has taken place.

6.7 Finite mixtures of random effects models

One can generalise the models in Section 5.2, which are mixtures of common age models, to finite mixtures of random effects models.

For example, we might consider a mixture of two components, each of the form described in Section 6.3, with parameters μ_1, σ_1 and μ_2, σ_2 and mixing proportions π_1 and π_2, where $\pi_1 + \pi_2 = 1$. But in so doing, we need to consider what *real* variation the parameters σ_1 and σ_2 represent. If they describe variation in true fission track ages between grains, what is it that distinguishes grains in one population from those in another? For ages obtained by the population method, such variation might be due to heterogeneity in uranium concentration between grains. But here we are considering the external detector method, where this variation is eliminated. Of course, if σ_1 and σ_2 describe an extra component of "experimental error" then, provided that they are known or can be estimated separately, they can be appropriately combined with σ_u before fitting mixtures as in Section 5.5.

We can imagine, for example, the two samples in Figure 6.1 pooled into a single sample, but here it is known which grain belongs to which population. So we need a scenario in which there is both heterogeneity and more than one distinct group in the same sample. For reasons of parsimony it would also be nice to be able to assume a common value of σ for each component, though this may not be realistic.

The above dilemma may be illustrated using the data from Mount Tom in Table 5.6. Suppose we fit finite mixtures as before, but adding an extra (known) common variance term σ^2 to each observation, so that the variance for grain u becomes $\sigma^2 + \sigma_u^2$, where σ_u is as in Section 5.7. Using $\sigma = 0, 0.1$

and 0.2 for illustration, and fitting $k = 2, 3, 4$ and 5 components, gives the following values of the maximum log likelihood, L_{max}:

k	$\sigma = 0$	$\sigma = 0.1$	$\sigma = 0.2$
2	23.31	51.73	84.89
3	62.21	78.01	96.14
4	77.19	84.24	96.20
5	77.34	84.24	96.20

When $\sigma = 0$, the values of L_{max} are the same as those obtained in Section 5.7 and suggest that $k = 4$ components are appropriate. When $\sigma = 0.1$, L_{max} has increased for every k, though $k = 4$ is still indicated. When $\sigma = 0.2$, again L_{max} increases, but now only $k = 3$ components are indicated. For each fixed σ, L_{max} of course increases with k, though insignificantly after sufficient numbers of components are included. But for each fixed k, L_{max} also increases with σ. If we increased σ further we would find that even fewer components were needed to explain the data, until a point came where the variation between grains was well explained by a single component with a large σ.

The upshot of this is that we really need to know the probable value(s) of σ, or perhaps of some of the other parameters, in order to find an appropriate model. Most importantly, we need to be clear about what variation σ is supposed to represent. In the next section we discuss a model that is essentially a mixture of one common age model and one random effects model, where the latter has a truncated distribution.

6.8 A minimum age model

The random effects model in Section 6.3 can provide a useful description of data when there is over-dispersion — for example to compare samples down a bore-hole or over an area. However, as with any model, it will not always be appropriate and its parameters will not always be geologically meaningful. In particular, if some grains in the sample have, at some stage in their history, been completely annealed (e.g., the fluorapatites) while others have not, then it is usually of more interest to estimate the common age of these grains rather than the central age of all grains, as the former would indicate the time since the sample was cooled.

For example, consider the thermal history (d) in Figure 1.3 in Chapter 1. In this scenario, fission tracks formed before t_u experience a maximum temperature of T_u, while those formed after t_u only experience a maximum temperature of T_0. Because all tracks spend some time near their maximum temperatures the distribution of lengths within a grain is effectively a two-component mixture with mean $(1 - p)\mu_{s0} + p\mu_{s1}$. The component mean lengths μ_{s0} and μ_{s1}, where μ_{s0} is greater than μ_{s1}, reflect the temperatures T_0 and T_u, while p is the proportion of tracks formed before t_u.

Now if there is compositional variation, μ_{s0} and μ_{s1} will typically vary between grains and hence so will ρ_s/ρ_i. Furthermore if T_0 is sufficiently low, μ_{s0} will not vary much between grains. If also T_u is sufficiently high to anneal tracks completely in some grains (those that are more sensitive to heat) then these grains will have a common ρ_s/ρ_i derived from tracks formed after t_u, with mean length μ_{s0}. Hence they will have a common fission track age t_u. For the remaining grains, which still contain tracks formed before t_u, the mean lengths $(1 - p)\mu_{s0} + p\mu_{s1}$ will be shorter than μ_{s0} and will vary, and hence so will ρ_s/ρ_i. Consequently, the ages t given by equation (3.10) will be older than t_u and will vary between grains.

The *minimum age* model is designed to be suitable for estimating t_u from fission track counts in such situations — i.e., for outcrop samples or cool samples that have experienced rapid uplift and erosion at some stage. Special cases of this model include the two-age model (Section 5.2) and random effects model (Section 6.3), so it also provides a statistical basis for assessing departures from these simpler models.

6.9 Data and statistical model with binomial errors

As usual, let the data be spontaneous and induced fission track counts for n grains, obtained by the external detector method in a single irradiation and counted by a single analyst. Denote these by N_{su} and N_{iu} for $u = 1, 2, \ldots, n$ and let $y_u = N_{su}$ and $m_u = N_{su} + N_{iu}$.

In the minimum age model, the values of $\beta = \log(\rho_s/\rho_i)$ for the sample grains are randomly drawn from a population in which a proportion π_γ have the same value γ and the remaining proportion $1 - \pi_\gamma$ are from a Normal distribution with mean μ and standard deviation σ that is truncated below at γ. So β either equals γ or is greater than γ. Thus, for grain u, β_u is drawn from this population and then the spontaneous track count y_u is drawn from a binomial distribution with index m_u and parameter $\theta_u = e^{\beta_u}/(1 + e^{\beta_u})$. As usual, the count y_u is observed, but the value of β_u is not.

Figure 6.3 illustrates the population distribution of β when γ is less than μ (left panel) and in the special case when $\gamma = \mu$ (right panel). In the former

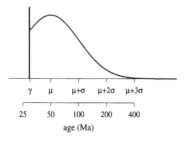
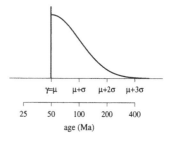

Figure 6.3 *Example population distributions for the four-parameter minimum age model (left panel) and the three-parameter model with $\gamma = \mu$ (right panel). The vertical bar located at γ indicates that there is a lump of probability π_γ at this value and the curve indicates the relative frequencies of values of $\beta = \log(\rho_s/\rho_i)$.*

case there are four parameters: μ, σ, γ and π_γ. The first two have a similar interpretation to μ and σ in the random effects model, corresponding to the central age t_μ (obtained by substituting $\rho_s/\rho_i = e^\mu$ in the fission track age equation) and the age dispersion in the non-truncated population. The parameter π_γ is the proportion of grains in the population that were completely annealed and γ can be interpreted on the age scale as the common age t_γ of these grains, corresponding to $\rho_s/\rho_i = e^\gamma$ in the fission track age equation. The model can thus be interpreted in terms of t_γ, t_μ, σ and π_γ.

In Figure 6.3 the age scale is approximately logarithmic, so the distribution of t has a lump at t_γ and is then positively skewed. In the left panel t_γ is about 30 Ma and t_μ is 50 Ma, while in the right panel t_γ and t_μ both equal 50 Ma. There are several sub-models of interest:

(a) When $\gamma = \mu$ the model reduces to a three-parameter model in which the truncated distribution of β is Half-Normal. Experience suggests that this is a useful parsimonious model for small numbers of grains.

(b) When σ approaches zero and $\gamma < \mu$ the model reduces to the two-age model in Section 5.2 with $t_1 = t_\gamma$, $t_2 = t_\mu$ and $\pi_1 = \pi_\gamma$.

(c) When γ approaches $-\infty$ (i.e., the minimum age approaches zero) and π_γ approaches zero the model reduces back to the random effects model of Section 6.3.

(d) Finally, the one parameter model in which all crystals have the same age arises when the variance of β approaches zero. This can occur in three ways: as π_γ approaches 1; as σ approaches 0 and either $\gamma \geq \mu$ or γ approaches μ; or as σ approaches 0, π_γ approaches 0 and $\gamma < \mu$, but γ does not approach μ.

6.10 Maximum likelihood estimation formulae

The log likelihood function L, as a function of μ, σ, π_γ and γ, is given by

$$L \;=\; \sum_{u=1}^{n} \log\left\{p(y_u|m_u)\right\} \;+\; \text{constant}\,,$$

with $p(y_u|m_u)$ given by

$$p(y_u|m_u) \;=\; \binom{m_u}{y_u}\left\{\frac{\pi_\gamma\, e^{\gamma y_u}}{(1+e^\gamma)^{m_u}} + \frac{1-\pi_\gamma}{1-\Phi\left(\frac{\gamma-\mu}{\sigma}\right)}\int_\gamma^\infty f_u(x)\,dx\right\}, \quad (6.10)$$

where

$$f_u(x) \;=\; \frac{e^{x y_u}}{(1+e^x)^{m_u}}\,\frac{1}{\sigma\sqrt{2\pi}}\,\exp\left\{-\frac{1}{2}\left(\frac{x-\mu}{\sigma}\right)^2\right\} \qquad (6.11)$$

and where

$$\Phi(z) = \int_{-\infty}^{z} \frac{1}{\sqrt{2\pi}} e^{-\frac{1}{2}x^2}\, dx \qquad (6.12)$$

is the standard Normal integral. To compute L for any given values of the parameters in general requires a separate numerical integration for each grain.

Maximum likelihood estimates $\hat{\mu}$, $\hat{\sigma}$, $\hat{\gamma}$ and $\hat{\pi}_\gamma$ are then obtained by numerical maximisation of L over μ, $\sigma \geq 0$, γ and $0 \leq \pi_\gamma \leq 1$. Their approximate covariance matrix may be obtained by inverting the 4×4 matrix of second derivatives of L, also calculated numerically. For the special case when $\gamma = \mu$ the maximisation is done with respect to the three parameters γ, σ and π_γ, and their covariance matrix is obtained by inverting the 3×3 matrix of second derivatives of L. Again, it may be more reliable to calculate confidence intervals from the profile log likelihood functions.

The statistical properties of this model, computational methods and properties of the maximum likelihood estimates obtained were discussed in detail by van der Touw *et al.* (1997). Properties of the estimates were studied there by fitting both correct and incorrect models to an extensive range of simulated data sets. Overall it was found that maximum likelihood estimation works well if there are sufficient data, at least as far as point estimates are concerned. Some conclusions from that study were:

- the four-parameter model needs a quite large number of grains and quite highly dispersed single grain ages to ensure that all of the parameters are well identified;

- the point estimate of the minimum age is reasonably good for mildly dispersed ages, but the other parameters are less well estimated;

- the 95% profile likelihood confidence interval for the minimum age is wide when the data are only mildly dispersed, and sometimes the lower confidence limit is misleading;

- setting $\mu = \gamma$ overcomes most of the above problems, although it creates a slight bias in the estimated minimum age;

- fixing $\pi_\gamma = 0$ in the estimation creates severe bias problems in all parameters when the true value is not zero.

A useful conclusion was that the minimum age parameter t_γ is often well identified even in situations where the other parameters in the model are not.

6.11 A minimum age model with Normal errors

As in Section 6.4, when the numbers of spontaneous and induced track per grain are reasonably large, simpler formulae can be obtained with z_u and σ_u given by (5.12) and (5.13). Now z_u is assumed to be from a Normal population with mean β_u and standard deviation σ_u, where β_u is either γ with probability π_γ or drawn from the truncated Normal population with parameters μ, σ and γ, with probability $1 - \pi_\gamma$. Then the probability density of z_u is

$$
f_u(z_u) = \frac{\pi_\gamma}{\sqrt{2\pi\sigma_u^2}} \exp\left\{ -\frac{(z_u - \gamma)^2}{2\sigma_u^2} \right\} +
$$

$$
\frac{1 - \pi_\gamma}{\sqrt{2\pi(\sigma^2 + \sigma_u^2)}} \frac{1 - \Phi\left(\frac{\gamma - \mu_{0u}}{\sigma_{0u}}\right)}{1 - \Phi\left(\frac{\gamma - \mu}{\sigma}\right)} \exp\left\{ -\frac{(z_u - \mu)^2}{2(\sigma^2 + \sigma_u^2)} \right\},
$$

where

$$\mu_{0u} = \frac{\mu/\sigma^2 + z_u/\sigma_u^2}{1/\sigma^2 + 1/\sigma_u^2} \quad \text{and} \quad \sigma_{0u} = \frac{1}{\sqrt{1/\sigma^2 + 1/\sigma_u^2}}$$

and $\Phi(z)$ is the standard Normal integral given by (6.12). Now in the log likelihood function, $p(y_u|m_u)$ given by (6.10) is replaced by $f_u(z_u)$. There are still no simple equations for calculating maximum likelihood estimates of μ, σ, γ and π_γ, but the log likelihood function is much easier to compute and again there are no numerical integrations needed. Of course, in other situations, where single grain ages are estimated directly, this Normal model is useful in its own right.

6.12 Example: apatite data from China

Let us apply the minimum age models to some data from the western Tarim basin in China. These data were kindly provided by Trevor Dumitru and their geological context is described in Sobel and Dumitru (1997). Here we look at three samples of apatite crystals, denoted by W_2, W_3 and W_4, taken from shallowly buried sandstones at neighbouring locations along a dry wash near the village of Wenguri. The counts are listed in Table 6.2.

The sand and apatite in these samples were derived from bedrock eroded off the Tian Shan ("Heavenly Mountains"). Major thrust faulting in the Tian

Table 6.2 *Spontaneous and induced fission track counts obtained by the external detector method for apatite grains from three locations W_2 (32 grains), W_3 (50 grains) and W_4 (40 grains) in the Western Tarim basin in China. Data from Sobel and Dumitru (1997).*

W_2				W_3						W_4					
N_s	N_i	N_s	N_i	N_s	N_i	N_s	N_i	N_s	N_i	N_s	N_i	N_s	N_i	N_s	N_i
8	133	32	197	13	220	55	110	19	181	24	194	4	66	5	46
17	136	5	131	42	114	4	64	7	34	11	49	8	125	29	113
19	210	17	304	2	7	4	24	36	280	31	41	12	107	11	115
9	146	16	374	80	229	5	45	9	65	40	133	72	146	7	99
7	58	29	405	11	104	3	152	17	131	6	121	24	298	30	91
25	257	39	482	7	54	9	59	5	57	33	60	36	217	2	76
2	35	26	369	12	176	16	117	9	22	17	221	26	263		
14	160	18	445	53	586	26	79	51	648	15	177	72	415		
6	109	20	240	14	182	6	69	41	482	52	80	20	230		
7	133	41	107	121	168	13	108	16	183	4	28	21	171		
7	35	26	146	12	188	29	328	51	720	17	103	26	160		
6	137	13	191	2	101	39	87	35	436	1	53	18	222		
16	209	25	364	91	458	65	79	28	273	23	115	15	176		
6	141	48	123	19	185	229	568	71	123	7	37	113	161		
16	190	19	471	5	12	133	323	45	120	5	49	27	237		
12	310			10	95	34	346	19	241	12	137	10	51		
18	175			48	595	32	195			5	55	62	145		

N_s = number of spontaneous tracks counted; N_i = number of induced tracks counted; $\hat{\zeta} = 361.5$ Ma $\times 10^{-6}$ cm^2; relative standard error of $\hat{\zeta} = 22.4/361.5 = 0.062$; $N_d = 6447$ tracks; $\hat{\rho}_d = 1.45 \times 10^6$ cm^{-2}.

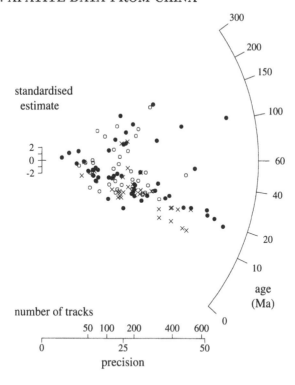

Figure 6.4 *Radial plot of fission track age estimates for three apatite samples in Table 6.2: W_2 (crosses), W_3 (closed circles), W_4 (open circles).*

Shan brought deeply buried rock rapidly to the earth's surface, where it was eroded, and the resulting sand and apatite grains (and other detrital minerals) were carried by rivers into the Tarim basin. The fission track age estimates are plotted in Figure 6.4. They vary over a wide range and there appears to be a fairly well-defined minimum value somewhere between 10 and 20 Ma.

Fission track ages may vary between crystals for a number of reasons. Here they are thought to reflect the differing provenance of the source material. Sobel and Dumitru (1997) noted that "the minimum age clusters are synchronous with or only slightly older than the depositional ages" suggesting that the grains were rapidly eroded in the provenance terrains and rapidly transported into the host sediment. The youngest ages are of particular interest as they may date the time of thrust faulting and major exhumation in the main sediment source areas, while the older crystals are from source areas where the deformation was less strong. Because the three samples are from neighbouring locations and their youngest ages appear to be comparable, it is of some interest to estimate a common youngest age for the pooled sample of 122 crystals.

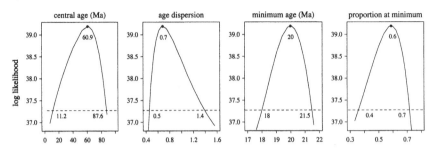

Figure 6.5 *Profile log likelihood functions for the four-parameter minimum age model fitted to the China data (Table 6.2).*

Another hypothesis suggested by Sobel and Dumitru (1997) is that, because of their geographical locations, the youngest ages should increase systematically in the order W_2 (youngest), W_3 (next youngest), W_4 (oldest). In Figure 6.4 there are two grains from sample W_3 and two from W_4 that have younger age estimates than any from W_2. Otherwise the lowest age clusters do look to be the the right order. These exceptions are the $(2, 101)$ and $(3, 152)$ grains in W_3 and $(1, 53)$ and $(2, 76)$ in W_4, so they all have small numbers of spontaneous tracks. We will use the model to look at this hypothesis more formally. Finally, because the variation in ages is thought to be mainly due to differing provenance, rather than the effects of heat and differing chemistry, we will also consider the use of discrete mixture models for these data.

Firstly, fitting the four-parameter minimum age model gives the following estimates and 95% confidence intervals:

minimum age (Ma)	$\hat{t}_\gamma = 20.0$	$(18.0 - 21.5)$
central age (Ma)	$\hat{t}_\mu = 60.9$	$(11.2 - 87.6)$
dispersion	$\hat{\sigma} = 0.70$	$(0.50 - 1.40)$
proportion at minimum	$\hat{\pi}_\gamma = 0.60$	$(0.40 - 0.70)$

The maximum log likelihood, where an arbitrary constant has been added, is $L_{\max} = 78.98$ and $BIC = -138.74$, where BIC is defined in (5.26). Profile log likelihood functions for t_μ, σ, t_γ and π_γ are shown in Figure 6.5 and indicate a reasonably well-behaved estimation procedure, with slightly non-symmetric confidence intervals. The estimates of σ and π_γ indicate a high dispersion, with a large proportion of crystals at the minimum age. The estimate of t_γ is 20.0 Ma, and is quite precise, with a 95% confidence interval from 18.0 to 21.5 Ma. But the 95% confidence interval for the central age t_μ is very wide and completely overlaps that of the minimum age t_γ, which possibly suggests some parameter redundancy.

It is therefore of worth trying the three-parameter model in which $\mu = \gamma$. This gives $L_{\max} = 77.52$ and $BIC = -140.63$ and the following estimates (and 95% confidence intervals):

minimum age (Ma)	$\hat{t}_\gamma = 19.1$	(16.3 – 20.8)
dispersion	$\hat{\sigma} = 1.20$	(1.00 – 1.50)
proportion at minimum	$\hat{\pi}_\gamma = 0.50$	(0.20 – 0.60)

The minimum age estimate is slightly lower than before, with a 95% confidence interval from 16.3 to 20.8 Ma, and the dispersion ($\sigma > 1$) is now very high.

Which model is preferred? The three-parameter model has a lower value of BIC and is therefore favoured by this criterion. The increase in L_{max} when going from three to four parameters can formally be tested for significance: we have $2 \times (78.98 - 77.52) = 2.92$ which gives a p-value of 0.085 from the upper tail of the $\chi^2(1)$ distribution. This indicates only weak evidence against the three-parameter model. Therefore the data do not really favour the four-parameter model, although the choice on statistical grounds is perhaps not clear cut.

When fitting models to the three samples separately, though, the four-parameter model is clearly over-parameterised, so we continue with the $\gamma = \mu$ model. To answer the question "Is the youngest age the same at each location?" we may fit the model, first with γ, σ and π_γ all separately estimated from each sample (nine parameters in all) and then with the same value of γ but separate σ and π_γ (seven parameters estimated). The difference in the two values of L_{max} is 5.38. So to test the hypothesis that γ is the same at each location we assess the significance of $2 \times 5.38 = 10.76$ from the χ^2-distribution with $9 - 7 = 2$ degrees of freedom. The upper tail probability gives a p-value of 0.005 which is indeed significant evidence that γ differs between locations. The three separate estimates of the youngest age (and 95% confidence intervals) in Ma are

location	$\hat{t}_\gamma$	95% CI
W_2	16.2	8.2 – 18.5
W_3	21.7	19.1 – 24.0
W_4	22.5	14.2 – 26.5

These estimates do increase systematically in the order hypothesised by Sobel and Dumitru (1997), although there is some overlap between the confidence intervals.

This example illustrates the power of parametric modelling and likelihood methods to extract non-obvious information from the data. A similar approach may be used to answer the question "Is the dispersion the same at each location?". This question is of less direct scientific interest, but for reasons of robustness it would be useful to know if it mattered whether we assumed different values of σ or a common value. Here we fit the model first with all parameters different and then with a common σ, so the difference in the number of parameters fitted is again $9 - 7 = 2$. The difference in L_{max} is now only 0.45, and $2 \times 0.45 = 0.90$ gives an upper tail p-value of 0.64 from the $\chi^2(2)$ distribution. This p-value is large and consistent with the hypothesis that σ is the same at each location.

Finally, let us look at the results of fitting discrete mixtures to each sample separately with a view to estimating the youngest age at each location. The table below shows values of $L_{\max}$ and BIC obtained for various numbers of ages (k) fitted to each sample separately. In addition, the last line gives the corresponding $L_{\max}$ and BIC for the three-parameter minimum age model.

	W_2		W_3		W_4	
k	$L_{\max}$	BIC	$L_{\max}$	BIC	$L_{\max}$	BIC
2	16.72	−23.04	16.84	−21.94	17.42	−23.77
3	24.41	−31.49	33.50	−47.44	32.73	−47.02
4	27.43	−30.60	35.60	−43.82	redundancy	
$\gamma = \mu$	22.49	−34.58	36.31	−60.88	32.35	−53.63

In every case attempting to fit more than four ages results in parameter re-dundancy. For all three samples the smallest BIC occurs when $k = 3$. For both samples W_3 and W_4 a model with $k = 3$ is clearly indicated, and the estimated youngest ages are, respectively, 22.6 Ma and 23.5 Ma. For sample W_2 a three-age mixture is also suggested with an estimated youngest age of 17.0 Ma. But also for W_2 the BIC value is very nearly as small for $k = 4$ and there is some slight evidence of the presence of an even younger component estimated at 11.9 Ma, with the next youngest at 20.1 Ma. This is no doubt due to the two youngest grains from sample W_4 (solid dots) that in Figure 6.4 look to be rather younger than the larger number near the 20 Ma radius.

Adopting a three-age model for each location, the estimates of the youngest age in Ma, with 95% confidence intervals, are:

location	$\hat{t}_1$	95% CI
W_2	17.0	15.1 – 18.8
W_3	22.6	20.6 – 24.7
W_4	23.5	20.3 – 26.7

These are comparable with those obtained from the minimum age model, though they are slightly higher and perhaps have less realistic confidence in-tervals. Note that the BIC values are all lower for the minimum age model, which has just three fitted parameters rather than five.

6.13 A synthetic mixture re-visited

It is instructive to look briefly at the result of fitting the four-parameter minimum age model to the data in Table 5.1. This is a synthetic mixture of two ages, which were successfully estimated in Section 5.3. Now we noted above that the two-age model is a special case of the four-parameter model with $\sigma = 0$, so in principle we should be able to recover the two-age model.

Indeed when the the four-parameter model is fitted to these data, the log likelihood $L_{\max}$ increases by just 0.49, compared with that for the two-age model, and the estimate of σ is 0.096 with a confidence interval that includes zero. The age estimates are $\hat{t}_\gamma = 235.4$ and $\hat{t}_\mu = 336.9$ and the estimate of π_γ is 0.38, which compare well with the estimates of t_1, t_2 and π_1 in Table 5.2. In other words, the program has effectively fitted the two-component mixture with practically the same results.

6.14 Grain age distributions

Mixture distributions often arise in provenance studies, where variation in true fission track ages reflects the differing source regions from which grains have been eroded, rather than (or in addition to) the effects of heat. The data from Mount Tom and China (Sections 5.7 and 6.12) are examples. Many provenance studies use fission tracks in zircon, rather than apatite, because they are less affected by moderate heating. In such studies it may be desired to look at a graph of the whole frequency distribution of true grain ages, or an estimate of it, before making specific inferences.

This is not straightforward to do. Firstly, the interpretation of the relative frequencies of different ages requires care, as they may reflect sampling processes and experimental constraints, rather than frequencies in nature. For example, for a discrete age mixture, the age estimates may truly reflect the ages of deposition of grains from different sources, but the mixing proportions might not reflect their relative abundance in nature — either because the sample collection in the field is not representative, or because the selection of grains within the sample is not.

Secondly, there are problems with interpreting a graph of the observed ages because their distribution is convoluted with a different measurement error distribution for each grain. Therefore the empirical distribution of observed ages can differ markedly from that of the true ages. Often the grain age distribution may be conveniently summarised by a discrete distribution, which may be estimated by the methods in Chapter 5. Then radial plots are effective in visualising both the single grain ages and the mixture component ages on one graph, as in Figure 5.3. For comparing data from several samples, it is likely to be more informative to plot the *fitted* distributions with respect to a common age scale, and to tabulate their summary statistics, than to plot observed distributions.

Sometimes it may be convenient to regard the underlying age distribution as being continuous and the same issues of interpretation arise here too. For a continuous mixture, suppose that we have fission track counts $y_u = N_{su}$ and $m_u = N_{su} + N_{iu}$ for grains $u = 1, 2, \ldots, n$, in the notation of Section 6.3. Then the conditional probability of y_u given m_u is

$$p(y_u|m_u) = \binom{m_u}{y_u} \int_{-\infty}^{\infty} e^{\beta y_u}[1 + e^{\beta}]^{-m_u} f(\beta)\, d\beta, \qquad (6.13)$$

where $f(\beta)$ is the probability density function of the log odds parameter β. This distribution may be transformed to a distribution of ages t using (6.1) where $\rho_s/\rho_i = e^{\beta}$ as usual. Each y_u is from a different distribution, but the same mixing distribution $f(\beta)$ is common to all grains. Equation (6.5) is of this form where $f(\beta)$ is Gaussian and is therefore of a specific shape. Other parametric forms for $f(\beta)$ could be assumed and their parameters estimated by maximum likelihood. While this approach is useful for estimating specific features of the age distribution, it may not be sufficiently flexible to reveal the shape of $f(\beta)$ in general.

In fact the amount of shape detail that it is possible to see is rather limited unless the binomial components of variation are small. It is well known that the "non-parametric maximum likelihood estimate" of $f(\beta)$ is necessarily discrete, and — more interestingly — is usually concentrated on a small number of points (Laird, 1978). For example, in a sample of 50 grains it may be concentrated on only three or four distinct values. So, unless some smoothing is applied, we are led back to looking at a discrete distribution. Goutis (1997) proposed a continuous estimate of $f(\beta)$ based on kernel density smoothing inside the integral in (6.13). This method is rather computer-intensive and is yet to be developed to ascertain its usefulness in practice.

6.15 Remarks

A question that often arises in fission track analysis is what to do when there is over-dispersion in a sample of single grain ages — in particular, when there is over-dispersion with respect to Poisson variation in N_s and N_i in the external detector method. Of course a sensible answer to this question must depend on the context. If the true density ratio ρ_s/ρ_i varies between grains, one needs to consider what possible forms such variation might take. The models discussed here and in Chapter 5 — finite mixtures, random effects and minimum age models — represent different forms of heterogeneity that might apply in different situations. Other models can be built, if necessary, from combinations or modifications of these, but it is important that the main parameters of interest represent meaningful quantities.

It may happen that apparent heterogeneity in ρ_s/ρ_i is partly due to further sources of "experimental error" or to artefacts of the data collection process, rather than "natural" variation. The above models can account for such features also, though every effort should be made to measure all experimental variation independently if possible. It is perhaps worth noting in passing that, if there is heterogeneity, the interpretation of some standard statistics may not be straightforward. For example, it is not clear what population quantity the pooled age represents if the grains do not have a common fission track age.

A general objective is to understand the pattern of single grain ages by building realistic models for the variation of true ρ_s/ρ_i values between grains. In the models considered here we think in terms of a notional population of grains whose ages belong to some distribution from which we are sampling. This population may sometimes reflect a real population in nature, or it may simply be a convenient way to describe variation about the parameters of interest.

6.16 Bibliographic notes

Parts of this chapter are based on Galbraith and Laslett (1993).

CHAPTER 7

Probability distributions of lengths and angles

The previous four chapters were concerned with statistical analysis and modelling of fission track counts. This chapter is about the theoretical statistical properties of track lengths and orientations. We derive the joint probability distributions of lengths and orientations of horizontal confined tracks, of semi-tracks intersecting a plane, and of their projections onto a plane, and discuss some implications.

We first consider isotropic length distributions, where track lengths do not depend on orientation. In this case, because tracks are randomly oriented, probability distributions relating to tracks intersecting a plane do not depend on which plane it is. From Section 7.5 on we consider anisotropic length distributions, where the plane of observation does matter, concentrating on the case where it is parallel to a prismatic face. All of these distributions, and others, may be derived from the main result of this chapter, equation (7.13), which gives a formula for the expected number of tracks that intersect a plane and that have some prescribed attribute.

To fix notation, Figure 7.1 shows three-dimensional coordinates, where the z-axis is parallel to the crystallographic c-axis and the yz-plane is parallel to a prismatic face. A track of length l intersects the yz-plane, with semi-track length t and projected semi-track length r. The angle between the projection of the track onto the observation plane and the c-axis is denoted by ω. Observers count tracks intersecting a chosen area of the plane, and they may also measure projected semi-track lengths r and the angle ω. It is possible, but harder, to measure the actual semi-track length t and its angle ϕ to the c-axis.

7.1 All tracks having the same length

Suppose that all tracks have exactly the same length l. The mean length μ equals l and the variance σ^2 is zero. Although this is highly idealised, it provides a useful intuitive basis for understanding the consequences of sampling by plane section. In the absence of variation in latent track lengths, any variation in semi-track lengths and projected semi-track lengths will be due to the random locations and orientations of tracks, along with the geometrical effects of truncation and projection.

Let random variables X_s and X_{ps} denote, respectively, the semi-track length and projected semi-track length of a track intersecting a plane. Because all

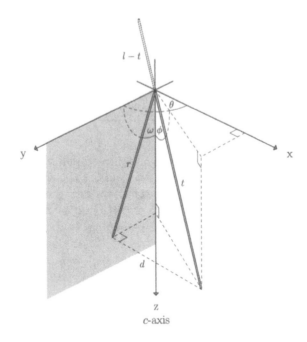

Figure 7.1 *Coordinate axes showing track length l, semi-track length t, projected semi-track length r, orientation angles ϕ and θ, and the angle ω between the projection of the track on the yz-plane and the z-axis. The yz-plane is the plane of observation. For apatite this is assumed to be a prismatic face, with the z-axis parallel to the crystallographic c-axis. The full length l cannot be observed because only the solid part is within the crystal, the dotted part having been polished away. (From Laslett et al., 1994.)*

tracks have the same length l and random orientation, the distributions of X_s and X_{ps} are the same for any plane. And because tracks can intersect the plane from a range of distances and at a variety of angles, semi-track lengths and projected semi-track lengths can both vary from 0 to l. We show below that the probability distribution of X_s is *uniform* over this range, with probability density function

$$f_s(t) \;=\; \frac{1}{l}, \qquad 0 \le t \le l \tag{7.1}$$

and mean

$$\mu_s = \mathrm{E}[X_s] = \tfrac{1}{2}l.$$

The variance of X_s is $l^2/12$ and the coefficient of variation (the ratio of the standard deviation to the mean) is therefore $l/\sqrt{12} \div l/2 = 1/\sqrt{3} \approx 0.58$, or 58%. In this case, the distribution (7.1) agrees with intuition: by symmetry, the plane of observation is equally likely to intersect a track and any point along its length.

Similarly we show that the projected semi-track length X_{ps} has a *triangular* distribution with probability density function

$$f_{ps}(r) = \frac{2}{l}\left(1 - \frac{r}{l}\right), \qquad 0 \leq r \leq l \qquad (7.2)$$

and mean

$$\mu_{ps} = \mathrm{E}[X_{ps}] = \tfrac{1}{3}l.$$

The variance of X_{ps} is $l^2/18$ and its coefficient of variation is therefore $l/\sqrt{18} \div l/3 = 1/\sqrt{2} \approx 0.71$, or 71%.

This probability density function has its maximum of $2/l$ at $r = 0$ and decreases linearly to zero at $r = l$. A short projected length r will arise from tracks either at a high angle to the plane or from atoms further away from the plane, so one can see intuitively why a large proportion of projected lengths must be relatively short. In particular, the operation of truncation halves the mean track length, and truncation followed by projection reduces the mean length to one third of its original value.

To derive (7.1) we calculate the theoretical proportion of tracks intersecting a unit area in the plane of observation whose semi-track lengths are greater than t, for any given t in the range $0 < t < l$. We do this by counting the expected number of such tracks and dividing by the corresponding expected number when $t = 0$. A similar argument is used to derive (7.2). These arguments follow from Cowan (1979, Lemma 1).

Without loss of generality, consider tracks intersecting the xy-plane in Figure 7.2 and projections onto that plane. It turns out that the calculations for the xy-plane are simpler than those for the yz-plane. Also, for reasons of symmetry we can restrict θ to the range $0 \leq \theta \leq \frac{1}{2}\pi$, so the joint probability density of (θ, ϕ) is

$$\tfrac{2}{\pi} \sin \phi, \qquad 0 \leq \theta \leq \tfrac{\pi}{2}, \quad 0 \leq \phi \leq \tfrac{\pi}{2}.$$

Figure 7.2 shows a parallelepiped with unit base area (shaded) that is parallel to the xy-plane and sides parallel to a track with orientation (θ, ϕ). Then, for $t < l$, the length of a semi-track intersecting a unit area in the xy-plane will be greater than t if (and only if) the bottom end of the track is in the parallelepiped. This parallelepiped has unit base area and height $(l - t) \cos \phi$, so it has volume

$$V = (l - t) \cos \phi.$$

Adapting the argument in Chapter 2, Section 2.3, we may regard the positions of the bottom ends of the tracks as a Poisson process with rate τ. So the expected number of tracks with their bottom ends in this parallelepiped is τV, and the expected number with orientation (θ, ϕ) is $\frac{2}{\pi} \sin \phi \, \tau V \, d\phi \, d\theta$. Adding these expected numbers for all orientations gives the expected number of semi-tracks with lengths greater than t, viz.,

$$E[N_s(t)] = \frac{2\tau}{\pi} \int_0^{\frac{\pi}{2}} \int_0^{\frac{\pi}{2}} (l - t) \cos \phi \sin \phi \, d\phi \, d\theta$$

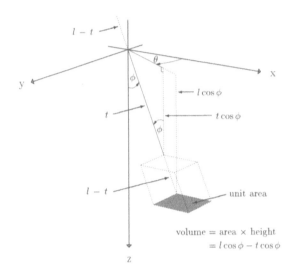

Figure 7.2 *The possible positions of a track of length l and orientation (θ, ϕ) that would intersect a unit area on the xy-plane and give rise to a semi-track length greater than t. This will happen if the bottom end of the track is in the parallelepiped with volume $(l - t)\cos\phi$. (From Laslett et al., 1994.)*

$$= \tfrac{1}{2}\tau\,(l - t).$$

Putting $t=0$ gives the expected total number of semi-tracks to be $\frac{1}{2}\tau l$, so the proportion with length greater than t is

$$P(X_s > t) = E[N_s(t)]/E[N_s(0)] = 1 - t/l, \qquad 0 \le t \le l.$$

Differentiating with respect to t and changing sign gives the uniform probability density function $f_s(t)$ in agreement with (7.1).

To derive (7.2), observe that we can only get a projected length greater than r if $l\sin\phi > r$, i.e., for

$$\arcsin(r/l) < \phi \le \tfrac{1}{2}\pi$$

and this will happen if the bottom end of the track is in the parallelepiped shown in Figure 7.2, where $r = t\sin\phi$. In terms of r, the volume of this parallelepiped is

$$V = (l - r/\sin\phi)\cos\phi$$

and by the same argument as before, the expected number of semi-tracks intersecting a unit area in the xy-plane that have projected lengths greater than r is

$$E[N_{ps}(r)] \quad = \quad \frac{2\tau}{\pi} \int_0^{\frac{\pi}{2}} \int_{\arcsin(r/l)}^{\frac{\pi}{2}} \left(l - \frac{r}{\sin\phi} \right) \cos\phi \, \sin\phi \, d\phi \, d\theta$$

$$= \tau \int_{\arcsin(r/l)}^{\frac{\pi}{2}} \cos\phi \, (l\sin\phi - r)\, d\phi\,.$$

Substituting $u = l\sin\phi - r$, $du = l\cos\phi \, d\phi$, gives

$$E[N_{ps}(r)] = \frac{\tau}{l} \int_0^{l-r} u\, du = \frac{\tau}{2l}(l-r)^2\,.$$

Putting $r = 0$ again gives the expected total number of semi-tracks to be $\frac{1}{2}\tau l$, so the proportion with projected lengths greater than r is

$$P(X_{ps} > r) = E[N_{ps}(r)]/E[N_{ps}(0)] = (1 - r/l)^2\,, \quad 0 \le r \le l\,.$$

Differentiating this with respect to r and changing the sign gives the probability density function $f_{ps}(r)$ in agreement with (7.2).

7.2 Each track having one of two lengths

A simple extension of the single-length model of the previous section is to suppose that there are just two possible latent track lengths l_0 and l_1. That is, independently of their orientation, a proportion p of tracks have length l_1 and a proportion $1 - p$ have length l_0, where $l_1 < l_0$. The mean track length is therefore

$$\mu = (1 - p)l_0 + pl_1\,.$$

Again this is an unrealistic model, but it serves to illustrate some consequences of varying track lengths, particularly the phenomenon of length-biased sampling.

To find the probability distributions of X_s and X_{ps}, consider the following argument. Let τ be the number of fissioned uranium atoms in a unit volume of the crystal. Then, on average, $(1 - p)\tau$ atoms produce tracks of length l_0 and $p\tau$ atoms produce tracks of length l_1. So, from the results for the single-length model, the expected total number of atoms producing tracks that intersect the surface is

$$\tfrac{1}{2}(1 - p)\tau l_0 + \tfrac{1}{2}p\tau l_1 = \tfrac{1}{2}\tau\mu\,.$$

Furthermore the expected number that produce tracks with semi-track length within dt of t is

$$\tfrac{1}{2}(1 - p)\tau l_0 \times f_{s0}(t)\, dt + \tfrac{1}{2}p\tau l_1 \times f_{s1}(t)\, dt\,,$$

where $f_{s0}(t)$ and $f_{s1}(t)$ are the probability density functions for semi-track lengths produced by tracks of length l_0 and l_1, respectively. That is, $f_{s0}(t)$ is uniform for $0 \le t \le l_0$ and $f_{s1}(t)$ is uniform for $0 \le t \le l_1$.

The overall relative frequency of semi-tracks of length t is therefore the ratio of these two quantities. Hence the probability density function of semi-track lengths (i.e., omitting the dt) is

$$f_s(t) = \frac{(1 - p)l_0}{\mu} f_{s0}(t) + \frac{pl_1}{\mu} f_{s1}(t)\,. \tag{7.3}$$

This is a mixture of the two component uniform distributions, f_{s0} and f_{s1}, with a proportion $(1 - p)l_0/\mu$ from the tracks of length l_0 and pl_1/μ from the tracks of length l_1. These proportions are *length biased*, so that relatively more of the semi-track lengths come from the longer tracks, compared with the original proportions $1 - p$ and p. The underlying reason for this is that, other things being equal, longer tracks are more likely than shorter ones to intersect the plane.

The mean semi-track length is

$$\mu_s = \frac{(1-p)l_0}{\mu} \times \frac{l_0}{2} + \frac{pl_1}{\mu} \times \frac{l_1}{2} = \frac{1}{2}\left(\mu + \frac{\sigma^2}{\mu^2}\right), \tag{7.4}$$

where $\sigma^2 = p(1 - p)(l_0 - l_1)^2$ is the variance of the latent track length distribution. The mean semi-track length is thus greater than $\frac{1}{2}\mu$, the mean for the single-length model, again as a consequence of length-biased sampling of tracks intersecting the plane.

By exactly the same argument, the probability density function of *projected* semi-track lengths is

$$f_{ps}(r) = \frac{(1-p)l_0}{\mu} f_{ps0}(r) + \frac{pl_1}{\mu} f_{ps1}(r), \tag{7.5}$$

which is a mixture of the two component *triangular* distributions, $f_{ps0}(r)$ and $f_{ps0}(r)$, each of the form (7.2) with parameter l_0 and l_1, respectively. Again the mixing proportions are the length-biased proportions $(1 - p)l_0/\mu$ and pl_1/μ.

The mean projected semi-track length is

$$\mu_{ps} = \frac{(1-p)l_0}{\mu} \times \frac{l_0}{3} + \frac{pl_1}{\mu} \times \frac{l_1}{3} = \frac{1}{3}\left(\mu + \frac{\sigma^2}{\mu^2}\right), \tag{7.6}$$

which is greater than $\frac{1}{3}\mu$, the value for the single-length model, for the same reason as before.

This example introduces the principle of deriving distributions of various quantities by mixing, in length-biased proportions, the corresponding distributions for the single-length model. We now extend the above results to apply first to a mixture of several lengths and then to any isotropic distribution of track lengths.

7.3 Several different lengths

Consider a more general situation, in which each latent track has one of several lengths $l_1, l_2, \ldots$. Let $p(l_i)$ be the proportion that have length l_i and let $p(l)$ be the same at all orientations. The mean length is now

$$\mu = \sum_{l_i=0}^{\infty} l_i p(l_i)$$

and the length-biased proportion is $l_i p(l_i)/\mu$. We can write the sum to ∞ because $p(l)$ will be zero for l greater than the maximum of the lengths. By the argument leading to (7.3) the probability density of t is obtained by adding up rectangles of the form (7.1) in length-biased proportions, $viz.$,

$$f_s(t) = \sum_{l=t}^{\infty} \frac{l_i p(l_i)}{\mu} \frac{1}{l_i} = \frac{1}{\mu} \sum_{l_i=t}^{\infty} p(l_i). \qquad (7.7)$$

The lower limit of the summation is t because (7.1) is zero for $l < t$.

The same argument applies to projected semi-track lengths, where the probability density of r is obtained by adding up triangles of the form (7.2) in length-biased proportions, $viz.$,

$$f_{ps}(r) = \sum_{l=r}^{\infty} \frac{l_i p(l_i)}{\mu} \frac{2}{l_i} \left(1 - \frac{r}{l_i}\right) = \frac{2}{\mu} \sum_{l_i=r}^{\infty} p(l_i) \left(1 - \frac{r}{l_i}\right). \qquad (7.8)$$

Here the lower limit of the summation is r because (7.2) is zero for $l < r$.

7.4 A general isotropic length distribution

Suppose that the length distribution is isotropic, so that $f_\phi(l) = f(l)$ for all l and the joint distribution of length and orientation is given by (2.7). This is a realistic model for lengths and orientations of unannealed tracks.

By extending the argument in the previous section, the probability density function of semi-track lengths can be written as a mixture of uniform probability densities, $1/l$, $0 < t < l$, one for each value of l, with length-biased mixing proportions $l f(l) dl / \mu$. That is,

$$f_s(t) = \int_t^{\infty} \left(\frac{1}{l}\right) \left(\frac{l f(l) \, dl}{\mu}\right), \qquad 0 < t < \infty,$$

which can be re-expressed as

$$f_s(t) = \frac{1}{\mu} \int_t^{\infty} f(l) \, dl = \frac{1}{\mu} \left(1 - F(t)\right), \qquad 0 < t < \infty, \qquad (7.9)$$

where $F(t) = \int_0^t f(l) dl$ is the cumulative distribution function of track length.

Equation (7.9) was originally derived by Laslett $et\ al.$ (1982). Because $f_s(t)$ is a weighted average of uniform probability densities on $(0, l)$ for different l, it follows from this construction that $f_s(t)$ must always be at its maximum value when $t = 0$ and cannot increase as t increases, regardless of the shape of $f(l)$. Note that $f_s(0) = 1/\mu$ so this maximum is determined by the mean track length μ.

The mean semi-track length is

$$\mu_s = \frac{1}{2} \left(\mu + \frac{\sigma^2}{\mu}\right), \qquad (7.10)$$

where σ^2 is the variance of the length distribution $f(l)$. Thus if track lengths do vary ($\sigma > 0$) then $\mu_s > \frac{1}{2}\mu$, essentially because the lengths are added in length-biased proportions, so more weight is given to longer lengths.

By the same argument, the projected semi-track length X_{ps} has probability density function given by

$$f_{ps}(r) = \int_r^\infty \frac{2}{l}\left(1 - \frac{r}{l}\right)\left(\frac{lf(l)\,dl}{\mu}\right), \qquad 0 < r < \infty,$$

which is a mixture of *triangular* density functions (one for each l) combined in length-biased proportions. This may be re-written as

$$f_{ps}(r) = \frac{2}{\mu}\int_r^\infty \left(1 - \frac{r}{l}\right) f(l)\,dl, \qquad 0 < r < \infty. \tag{7.11}$$

Again, it follows from this construction that whatever shape $f(l)$ has, $f_{ps}(r)$ must always have its maximum at $r = 0$ and must decrease as r increases. The mean projected semi-track length is

$$\mu_{ps} = \mathrm{E}(X_{ps}) = \frac{1}{3}\left(\mu + \frac{\sigma^2}{\mu}\right), \tag{7.12}$$

which, if $\sigma > 0$, is greater than $\frac{1}{3}\mu$ for the same reason as before.

To illustrate the above formulae, Figure 7.3 shows $f_s(t)$ given by (7.9) and $f_{ps}(r)$ given by (7.11) for each of four distributions $f(l)$. In case (a), $f(l)$ has a fairly small coefficient of variation ($\approx 7\%$) and $f_s(t)$ and $f_{ps}(r)$ do not depart markedly from the idealised rectangular and triangular forms. In (b), $f(l)$ has a larger coefficient of variation ($\approx 18\%$) and there is more evidence of extra dispersion in $f_s(t)$ and $f_{ps}(r)$. In (c) and (d), the mean and standard deviation of l are the same, but the shape of $f(l)$ is very different — and would correspond to very different thermal histories. This difference is more subtle in $f_s(t)$ and barely noticeable in $f_{ps}(r)$. Some practical implications of this are noted in Chapter 8.

7.5 A general anisotropic length distribution

In the next sections we consider the anisotropic case, where the joint distribution of track lengths and orientations is given by (2.3), *viz.*,

$$f(l, \theta, \phi) = \tfrac{1}{2\pi}\sin\phi\, f_\phi(l), \qquad 0 \le \theta \le 2\pi, \quad 0 \le \phi \le \tfrac{\pi}{2} \quad 0 < l < \infty.$$

Here, expected numbers of tracks intersecting a plane, and hence the probability distributions of projected lengths and orientations, depend on the orientation of the plane of observation. Let ψ be the angle between this plane and the c-axis (see Chapter 2, Section 2.3). In principle, orientations are informative, in addition to lengths, so we derive distributions involving both.

We first generalise equations (2.9) and (2.15) in Chapter 2 to count tracks that have some specific attribute. Consider the scenario in Sections 2.2 and 2.3. Let A denote some attribute of interest and let $P(A\,|\,l, \phi, \theta)$ be the probability that a track of length l and orientation (θ, ϕ) has that attribute. For instance, A might denote the attribute that the track is longer than a given length, or that it is at a lower angle to the c-axis than a given angle, or both.

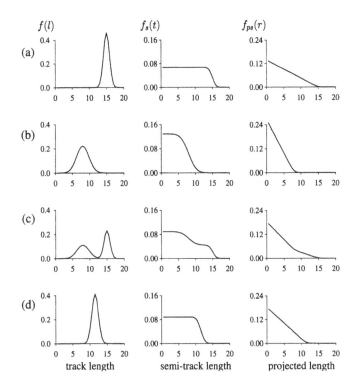

Figure 7.3 *Isotropic semi-track length distributions $f_s(t)$ and projected semi-track length distributions $f_{ps}(r)$ corresponding to four length distributions $f(l)$: (a) Normal with mean $\mu = 15$ and standard deviation $\sigma = 0.95$; (b) Normal with $\mu = 8$ and $\sigma = 3.96$; (c) a 50:50 mixture of (a) and (b); and (d) Normal with $\mu = 11.5$ and $\sigma = 1.42$.*

Let N_A denote the number of tracks that intersect a unit area in the plane of observation *and* that have the attribute A. Then from the argument in Section 2.2, N_A has a Poisson distribution with mean $E[N_A]$, say. Further, by the same argument as that leading to (2.9), but considering just those tracks with attribute A, the expected number that intersect the unit area in the plane of observation is

$$E[N_A] = \frac{\tau}{2\pi} \int_0^{2\pi} \int_0^{\frac{\pi}{2}} \sin\phi \int_0^\infty f_\phi(l)\, h(l,\phi,\theta,\psi)\, P(A\,|\,l,\phi,\theta)\, dl\, d\phi\, d\theta\,,$$

(7.13)

where $h(l,\phi,\theta,\psi)$ is given by equation (2.10), *viz.*,

$$h(l,\phi,\theta,\psi) = l\,|\sin\phi\cos\psi\cos\theta - \sin\psi\cos\phi\,|\,.$$

7.6 Distributions on a prismatic face

When the plane of observation is a prismatic face (i.e., $\psi = 0$), then

$$h(l, \phi, \theta, 0) = l \sin\phi \, |\cos\theta|$$

and (7.13) becomes

$$E[N_A] = \frac{\tau}{2\pi} \int_0^{2\pi} |\cos\theta| \int_0^{\frac{\pi}{2}} \sin^2\phi \int_0^\infty l \, f_\phi(l) \, P(A \,|\, l, \phi, \theta) \, dl \, d\phi \, d\theta. \quad (7.14)$$

By specifying different attributes A, equation (7.14) may be used to derive various distributions of interest for tracks intersecting a plane parallel to a prismatic face. We first consider the more complicated distributions, relating to projected semi-tracks, then use these results to deduce corresponding results for semi-tracks.

7.6.1 Projected semi-track lengths

Let the random variable X_{ps} denote the projected semi-track length of a track intersecting the observation plane (i.e., the yz-plane, parallel to a prismatic face). Let A be the event that $X_{ps} > r$ and let $N_A = N_{ps}(r)$ denote the number of tracks intersecting a unit area in this plane and whose projected semi-track lengths exceed r. Then the probability that X_{ps} exceeds r is given by

$$\Pr\{X_{ps} > r\} = E[N_{ps}(r)] \,/\, E[N_{ps}(0)]. \quad (7.15)$$

The right-hand side is the ratio of the expected number of tracks with the required property to the total expected number of tracks. As in Section 7.1, this argument follows from Cowan (1979, Lemma 1). We will use equation (7.14) to find $E[N_{ps}(r)]$ and hence find the distribution of X_{ps} from (7.15).

Referring to Figure 7.4, a track with semi-track length t has projected semi-track length $r = gt$, obtained by repeated application of Pythagoras' theorem, where

$$g = \sqrt{\cos^2\phi + \sin^2\phi \, \sin^2\theta}. \quad (7.16)$$

Also, the probability that a track with length l and orientation (ϕ, θ) has projected semi-track length greater than r is given by

$$P(A \,|\, l, \phi, \theta) = \begin{cases} (1 - r/lg) & \text{if } lg > r, \\ 0 & \text{if } lg \le r, \end{cases}$$

where A is the event $X_{ps} > r$. Then from (7.14),

$$\begin{aligned} E[N_{ps}(r)] &= \frac{\tau}{2\pi} \int_0^{2\pi} |\cos\theta| \int_0^{\frac{\pi}{2}} \sin^2\phi \int_{\frac{r}{g}}^\infty (1 - r/g) f_\phi(l) \, dl \, d\phi \, d\theta \\ &= \frac{2\tau}{\pi} \int_0^{\frac{\pi}{2}} \cos\theta \int_0^{\frac{\pi}{2}} \sin^2\phi \int_{\frac{r}{g}}^\infty (1 - r/g) f_\phi(l) \, dl \, d\phi \, d\theta \quad (7.17) \end{aligned}$$

since each angle θ, $\frac{\pi}{2} - \theta$, $\theta + \pi$, and $\frac{3\pi}{2} - \theta$ gives the same projected length

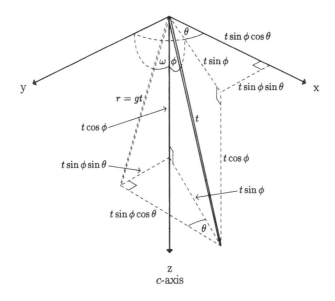

Figure 7.4 *Coordinate axes showing the projection* $r = gt$ *of a semi-track of length* t *on a prismatic face (the yz-plane). From Galbraith and Laslett (1988).*

(for fixed l and ϕ); or equivalently, each leaves both $|\cos\theta|$ and g unchanged. Putting $r = 0$ in (7.17) gives

$$\mathrm{E}[N] = \mathrm{E}[N_{ps}(0)] = \frac{2\tau}{\pi} \int_0^{\frac{\pi}{2}} \sin^2\phi\, \mathrm{E}[l|\phi]\, d\phi, \qquad (7.18)$$

which may also be written as $\rho = \frac{1}{2}\tau\mu^*$ where ρ is the track density and μ^* is the equivalent isotropic length given by

$$\mu^* = \frac{4}{\pi} \int_0^{\frac{\pi}{2}} \sin^2\phi\, \mathrm{E}[l|\phi]\, d\phi \qquad (7.19)$$

as in equations (2.16) and (2.17) in Chapter 2. Thus from (7.15), (7.17) and (7.18),

$$\Pr\{X_{ps} > r\} = \frac{4}{\pi\mu^*} \int_0^{\frac{\pi}{2}} \cos\theta \int_0^{\frac{\pi}{2}} \sin^2\phi \int_{\frac{r}{g}}^{\infty} (l - r/g) f_\phi(l)\, dl\, d\phi\, d\theta, \quad (7.20)$$

where g is given by (7.16). Differentiating with respect to r and changing the sign gives the probability density function $f_{ps}(r)$ of projected semi-track lengths:

$$f_{ps}(r) = \frac{4}{\pi\mu^*} \int_0^{\frac{\pi}{2}} \cos\theta \int_0^{\frac{\pi}{2}} \frac{\sin^2\phi}{g}\left(1 - F_\phi(r/g)\right) d\phi\, d\theta, \qquad 0 < r < \infty,$$

$$(7.21)$$

where

$$F_\phi(x) = \int_0^x f_\phi(l) \, dl$$

is the (cumulative) distribution function of lengths of tracks at angle ϕ to the c-axis.

Moments of the distribution of X_{ps} can be found conveniently from (7.20) using the results that, for a positive continuous random variable X,

$$E(X) = \int_0^\infty \Pr(X > x) \, dx, \qquad E(X^2) = \int_0^\infty 2x \Pr(X > x) \, dx$$

and so on. The standard deviation of the distribution of X can be obtained from

$$\mathrm{sd}(X) = \sqrt{\mathrm{var}(X)} = \sqrt{E(X^2) - [E(X)]^2} \,. \tag{7.22}$$

For example, the mean projected semi-track length μ_{ps} is

$$\mu_{ps} = E[X_{ps}] = \int_0^\infty \Pr\{X_{ps} > r\} \, dr \,.$$

We may evaluate this by first integrating the inner integral in (7.20) over r, which gives

$$\int_0^\infty \int_{\frac{r}{g}}^\infty (1 - r/g) f_\phi(l) \, dl \, dr = \tfrac{1}{2} \int_0^\infty g l^2 f_\phi(l) \, dl = \tfrac{1}{2} g \, E[l^2|\phi] \,.$$

Then, substituting for g from (7.16),

$$\mu_{ps} = \frac{2}{\pi \mu^*} \int_0^{\frac{\pi}{2}} \int_0^{\frac{\pi}{2}} \sin^2\phi \, \cos\theta \left(\cos^2\phi + \sin^2\phi \, \sin^2\theta \right)^{\frac{1}{2}} E[l^2|\phi] \, d\phi \, d\theta \,.$$

As $E[l^2|\phi]$ does not depend on θ, this may be reduced, using standard integrals, to

$$\mu_{ps} = \frac{1}{3\mu^*} \frac{3}{\pi} \int_0^{\frac{\pi}{2}} \left(\sin^2\phi + \sin\phi \, \cos^2\phi \, \log(\sec\phi + \tan\phi) \right) E[l^2|\phi] \, d\phi \,. \tag{7.23}$$

Except for the factor $1/3\mu^*$ this is a weighted average of $E[l^2|\phi]$ over values of ϕ. Compare this equation with (7.12) in the isotropic case, where the factor $1/3\mu^*$ becomes $1/3\mu$ and the remaining part reduces to $E[l^2] = \mu^2 + \sigma^2$.

Higher moments may be found similarly. In particular,

$$E[X_{ps}^2] = 2 \int_0^\infty r \Pr\{X_{ps} > r\} \, dr$$

$$= \frac{1}{6\mu^*} \frac{8}{\pi} \int_0^{\frac{\pi}{2}} \left(\sin^2\phi \cos^2\phi + \tfrac{1}{3} \sin^4\phi \right) E[l^3|\phi] \, d\phi \tag{7.24}$$

after substituting (7.20) and simplifying. This is $1/6\mu^*$ times a weighted average of $E[l^3|\phi]$ over ϕ. The standard deviation of X_{ps} may be obtained from (7.23) and (7.24) using (7.22).

7.6.2 Projected semi-track lengths and angles

In a similar way, by specifying the appropriate form for $P(A \mid l, \phi, \theta)$ in (7.13), we can obtain the joint probability density function $f_{ps}(r, \omega)$ of the projected length r and the angle ω between the projected semi-track and the c-axis (the z axis in Figure 7.1). Because of anisotropy of the length distribution, $f_{ps}(r, \omega)$ should depend on ω, and it is of interest to quantify this dependence.

Let the random variables X_{ps} and W denote, respectively, the projected semi-track length of a track intersecting the observation plane and the angle between this projection and the c-axis (i.e., the random variables corresponding to r and ω). Consider the event A that $X_{ps} > r$ and $W \le \omega_0$. The joint distribution of X_{ps} and W can be obtained from

$$P(X_{ps} > r, W \le \omega_0) = E[N_A] / E[N].$$

The probability that a track with length l and orientation (ϕ, θ) has projected semi-track length greater than r and projected angle to the c-axis less than or equal to ω is

$$P(A \mid l, \phi, \theta) = \begin{cases} (1 - r/lg) & \text{if } lg > r \text{ and } \tan\phi \sin\theta \le \tan\omega, \\ 0 & \text{otherwise.} \end{cases}$$

Using this expression in equation (7.14) and transforming (θ, ϕ) to (ω, ϕ) gives, after some algebra, the following expression for $P(X_{ps} > r, W \le \omega_0)$:

$$\frac{4}{\pi\mu^*} \int_0^{\omega_0} \sec^2\omega \, d\omega \int_\omega^{\frac{\pi}{2}} \sin\phi \cos\phi \, d\phi \int_{\frac{r\cos\omega}{\cos\phi}}^{\infty} \left(1 - \frac{r\cos\omega}{\cos\phi}\right) f_\phi(l) \, dl. \qquad (7.25)$$

Differentiating with respect to ω_0, and then writing ω in place of ω_0 leads to

$$\frac{4}{\pi\mu^*} \sec^2\omega \int_\omega^{\frac{\pi}{2}} \sin\phi \cos\phi \, d\phi \int_{\frac{r\cos\omega}{\cos\phi}}^{\infty} \left(1 - \frac{r\cos\omega}{\cos\phi}\right) f_\phi(l) \, dl. \qquad (7.26)$$

Then differentiating with respect to r (and changing sign) gives the joint probability density function of X_{ps} and W, viz.,

$$f_{ps}(r, \omega) = \frac{4}{\pi\mu^*} \sec\omega \int_\omega^{\frac{\pi}{2}} \sin\phi \left\{ 1 - F_\phi\left(\frac{r\cos\omega}{\cos\phi}\right) \right\} d\phi, \qquad (7.27)$$

for $0 < r < \infty$ and $0 < \omega < \frac{\pi}{2}$.

The marginal probability density function of projected semi-track angles ω can be obtained either by integrating $f_{ps}(r, \omega)$ over all r, or by putting $r = 0$ in (7.26) to give

$$f_{ps}(\omega) = \frac{4}{\pi\mu^*} \sec^2\omega \int_\omega^{\frac{\pi}{2}} \sin\phi \cos\phi \, E[l \mid \phi] \, d\phi, \qquad 0 < \omega < \frac{\pi}{2}. \qquad (7.28)$$

If $E[l \mid \phi]$ is constant, this reduces to a uniform distribution for ω between 0 and $\frac{2}{\pi}$. But if $E[l \mid \phi]$ varies with ϕ then $f_{ps}(\omega)$ will vary with ω. Thus in principle, strong anisotropy of the track length distribution should be evident, not only in the distribution of r but also in that of ω.

7.6.3 Projected semi-track lengths at a given angle

For completeness, we note here formulae for the conditional distribution of X_{ps} given $W = \omega$, which can easily be obtained from the previous sub-section. The probability density function is

$$f_{ps}(r|\omega) = \frac{f_{ps}(r,\omega)}{f_{ps}(\omega)},$$

where $f_{ps}(r,\omega)$ and $f_{ps}(\omega)$ are given by (7.27) and (7.28).

The mean of this distribution, following our earlier method, can be found by integrating $P(X_{ps} > r|\omega)$, given by (7.26), over $0 < r < \infty$, and can be expressed as

$$E(X_{ps}|\omega) = \frac{\frac{1}{2}\sec\omega \int_\omega^{\frac{\pi}{2}} \sin\phi \cos^2\phi\, E[l^2|\phi]\, d\phi}{\int_\omega^{\frac{\pi}{2}} \sin\phi \cos\phi\, E[l|\phi]\, d\phi}. \qquad (7.29)$$

Similarly, the second moment is

$$E(X_{ps}^2|\omega) = \frac{\frac{1}{3}\sec^2\omega \int_\omega^{\frac{\pi}{2}} \sin\phi \cos^3\phi\, E[l^3|\phi]\, d\phi}{\int_\omega^{\frac{\pi}{2}} \sin\phi \cos\phi\, E[l|\phi]\, d\phi}. \qquad (7.30)$$

7.6.4 Semi-track lengths

Let the random variable X_s denote the actual semi-track length of a track intersecting the observation plane (Figure 7.4) and let $N_s(t)$ denote the number of tracks intersecting a unit area in this plane and whose semi-track lengths exceed t. Then, by the same argument that led to (7.15), the probability that X_s exceeds t is given by

$$\Pr\{X_s > t\} = E[N_s(t)] / E[N_s(0)]$$

and the derivation of the distribution of X_s follows the same lines as that for X_{ps}. In fact, there is no need to repeat the argument as we can deduce the distribution of X_s simply by putting $g = 1$ and $r = t$ in (7.20). Hence, after integrating over θ,

$$\Pr\{X_s > t\} = \frac{4}{\pi\mu^*} \int_0^{\frac{\pi}{2}} \sin^2\phi \int_t^\infty (l-t) f_\phi(l)\, dl\, d\phi, \qquad (7.31)$$

with corresponding probability density

$$f_s(t) = \frac{4}{\pi\mu^*} \int_0^{\frac{\pi}{2}} \sin^2\phi\, \{1 - F_\phi(t)\}\, d\phi. \qquad (7.32)$$

Note that $f_s(t)$ is proportional to a weighted average over ϕ of the monotone decreasing functions $1 - F_\phi(t)$, so is still monotone decreasing with increasing t. Some examples are plotted in Figure 7.5.

Moments of this distribution can be found in the same way as before. The mean semi-track length μ_s is

$$\mu_s = E[X_s] = \frac{1}{2\mu^*} \frac{4}{\pi} \int_0^{\frac{\pi}{2}} \sin^2\phi \, E[l^2|\phi] \, d\phi, \qquad (7.33)$$

which is $1/2\mu^*$ times a weighted average of $E[l^2|\phi]$ with weight $\sin^2\phi$. Again, in the isotropic case this reduces to $1/2\mu$ times $E[l^2] = \mu^2 + \sigma^2$ in agreement with (7.10). Likewise, the second moment is

$$E[X_s^2] = \frac{1}{3\mu^*} \frac{4}{\pi} \int_0^{\frac{\pi}{2}} \sin^2\phi \, E[l^3|\phi] \, d\phi, \qquad (7.34)$$

from which the standard deviation of X_s can be obtained.

The formulae for the distribution and moments of semi-track lengths t are understandably somewhat simpler than those for their projections r, these lengths having been truncated but not projected. In principle, measurements of t are more informative than measurements of r (see for example Figures 7.3 and 7.5) but in practice it is much easier to measure r.

7.6.5 Illustrative probability density functions

It is hard to appreciate the general form of $f_{ps}(r)$ given by the rather complicated formula (7.21). Figure 7.5 illustrates some anisotropic distributions of projected semi-track lengths along with distributions of full lengths and of semi-track lengths. These have been calculated using a parametric model described in Section 7.11. Here $f_\phi(l)$ is Normal with mean $E[l|\phi] = \eta + \xi \cos\phi$ and standard deviation 0.85, where η and ξ are given in terms of μ by (7.53). Then $f(l)$ is obtained from equation (2.4), $f_{ps}(r)$ from (7.21) and $f_s(t)$ from (7.32). The differences from the isotropic case, Figure 7.3, are quite subtle, the most obvious one being in the length distribution $f(l)$, which is more platykurtic in the anisotropic case.

As for the isotropic case, the shape of $f_{ps}(r)$ is monotone decreasing with increasing r, regardless of the shape of the underlying distribution $f_\phi(l)$ of track lengths at angle ϕ. A consequence is that it is inherently difficult to distinguish between different thermal histories by looking at observed distributions of projected semi-track lengths. For example, the distribution $f_{ps}(r)$ for a thermal history giving rise to a unimodal $f_\phi(l)$ may not look very different from that for a very different thermal history leading to a bimodal $f_\phi(l)$. However, it should be noted that more information may be obtained from the joint distribution of lengths *and* angles.

7.7 Horizontal confined track lengths

It sometimes happens that a fission track may be etched out and the full track seen, even though it has not intersected the etched surface of the crystal. This can happen if it intersects a crack in the crystal down which the etchant has

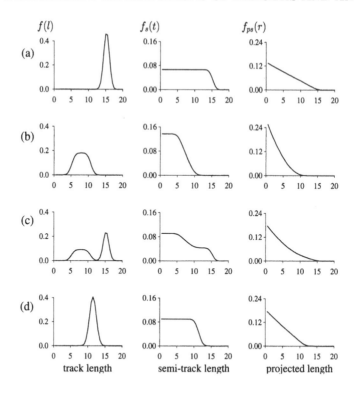

Figure 7.5 *Semi-track length distributions $f_s(t)$ and projected semi-track length distributions $f_{ps}(r)$ corresponding to four anisotropic length distributions $f(l)$: (a) with mean $\mu = 15$ and standard deviation $\sigma = 0.87$; (b) with $\mu = 8$ and $\sigma = 1.79$; (c) a 50:50 mixture of (a) and (b); and (d) with $\mu = 11.5$ and $\sigma = 0.97$.*

passed, or if it intersects another track that has been etched (i.e., a semi-track) which again provides a pathway for the etchant. Examples of such *confined* tracks may be seen in Figure 1.2. One that is etched via a crack or cleavage in the crystal is called a *tincle* (*track-in-cleavage*) and one etched via another track is called a *tint* (*track-in-track*). The discovery that some fully confined tracks could be etched and measured was a major advance in fission track analysis, because the full lengths of tracks could be measured, without truncation or projection, to give a much more informative estimate of the length distribution than that obtained by measuring r or even t.

In practice, for several reasons that are mentioned in Chapter 8, an observer usually selects those confined tracks that are *horizontal*, i.e., parallel to the plane of observation, and will effectively measure l and ϕ directly for these. Because the track is parallel to the plane, l equals its projection onto the plane, so the observer simply identifies the (y, z) co-ordinates (in the plane) of the ends the the track and hence calculates l. The angle ϕ is calculated by identifying, in addition, the (y, z) coordinates of two points aligned in the direction of the c-axis. It is customary to restrict ϕ to the range from 0 to $\frac{\pi}{2}$ for definiteness.

A *horizontal confined track* is thus one that is parallel to the etched surface (the yz-plane in Figure 7.1) but that has not intersected it, so that no part of the track has been polished away. The possible orientations of such a track correspond to a strip of infinitesimal constant width around a great circle on the surface of the unit sphere. Since the orientation of tracks is uniform on the sphere, the angle of horizontal confined tracks to the c-axis must be uniformly distributed, and so ϕ must be uniformly distributed between 0 and $\frac{\pi}{2}$. Hence the theoretical probability density function of the horizontal confined track length X_{hc} is

$$f_{hc}(l) \; = \; \frac{2}{\pi} \int_0^{\frac{\pi}{2}} f_\phi(l) \, d\phi. \tag{7.35}$$

The mean and second moment of X_{hc} are

$$E[X_{hc}] \; = \; \frac{2}{\pi} \int_0^{\frac{\pi}{2}} E[l|\phi] \, d\phi, \quad E[X_{hc}^2] \; = \; \frac{2}{\pi} \int_0^{\frac{\pi}{2}} E[l^2|\phi] \, d\phi \tag{7.36}$$

and similarly for higher moments. In other words, the moments of X_{hc} are the corresponding conditional moments of l given ϕ averaged with respect to a uniform distribution of ϕ.

Lengths and orientations of horizontal confined tracks are much more informative than are those for semi-tracks and projected semi-tracks, and are therefore more useful for thermal history analysis. In practice, though, their measurement involves a number of sampling and observational biases, and a true random sample of such tracks is virtually never achieved. So the distribution of *measured* lengths will depart from that given in (7.35). These biases include length bias, orientation bias and others to be discussed in Chapter 8.

7.8 Some explicit formulae

Nearly all of the integrals in the previous sections can be evaluated explicitly if the moments of $f_\phi(l)$ are polynomials in $\cos \phi$. Because smooth functions can generally be well-approximated by low-order polynomials this provides a simple method of approximating the various formulae, which can also be used to gain some insight into their behaviour.

An efficient method of evaluating the integrals is to use the formula

$$\int_0^{\frac{\pi}{2}} \sin^i \phi \cos^j \phi \, d\phi \; = \; \tfrac{1}{2} B\left(\tfrac{i+1}{2}, \tfrac{j+1}{2}\right) \; = \; \tfrac{1}{2} \Gamma\left(\tfrac{i+1}{2}\right) \Gamma\left(\tfrac{j+1}{2}\right) / \Gamma\left(\tfrac{i+j+2}{2}\right),$$

where B and Γ denote the Beta and Gamma functions. These are related by

$$B(x, y) = \frac{\Gamma(x)\Gamma(y)}{\Gamma(x+y)},$$

where the Gamma function values for $x \geq \frac{1}{2}$ can be found from

$$\Gamma(0) = \Gamma(1) = 1, \quad \Gamma(\tfrac{1}{2}) = \sqrt{\pi}, \quad \text{and} \quad \Gamma(x+1) = x\Gamma(x).$$

For example, to evaluate μ given by equation (2.5) one needs to integrate $\sin \phi$, $\sin \phi \cos \phi$, $\sin \phi \cos^2 \phi$, $\sin \phi \cos^3 \phi$, and so on, between 0 and $\frac{\pi}{2}$. By the above formula, these integrals are, respectively,

$$\tfrac{1}{2}B(1,\tfrac{1}{2}) = 1, \quad \tfrac{1}{2}B(1,1) = \tfrac{1}{2}, \quad \tfrac{1}{2}B(1,\tfrac{3}{2}) = \tfrac{1}{3}, \quad \tfrac{1}{2}B(1,2) = \tfrac{1}{4}$$

and so on.

For simplicity, suppose that $E[l|\phi]$ is closely approximated by a quadratic in $\cos \phi$, i.e., let

$$E[l|\phi] = \mu_0 + \mu_1 \cos \phi + \mu_2 \cos^2 \phi \tag{7.37}$$

for some suitable constants μ_0, μ_1 and μ_2. Further powers of $\cos \phi$ could be added if necessary, but experience with empirical measurements suggests that (7.37) may be good enough. Some specific parametric forms for $E[l|\phi]$ are mentioned briefly later (see Section 7.11) but here we are simply using (7.37) as a method of evaluating the formulae in this chapter. For higher moments we will use the notation

$$E[l^2|\phi] = \gamma_0 + \gamma_1 \cos \phi + \gamma_2 \cos^2 \phi + \cdots + \gamma_k \cos^k \phi \tag{7.38}$$

and

$$E[l^3|\phi] = \delta_0 + \delta_1 \cos \phi + \delta_2 \cos^2 \phi + \cdots + \delta_k \cos^k \phi \tag{7.39}$$

using higher powers of $\cos \phi$ if necessary.

7.8.1 Mean length and track density

From equation (2.5) for the mean track length μ, substituting (7.37) and using the above results gives

$$\mu = \mu_0 + \tfrac{1}{2}\mu_1 + \tfrac{1}{3}\mu_2. \tag{7.40}$$

Further, from equation (2.11) for the density of tracks intersecting an arbitrary plane, substituting for $E[l|\phi]$ from equation (7.37) it can be shown, after some algebra, that

$$\rho = \tfrac{1}{2}\tau\left(\mu_0 + \tfrac{4}{3\pi}(\cos\psi + \psi\sin\psi)\mu_1 + \tfrac{1}{4}(1 + \sin^2\psi)\mu_2\right). \tag{7.41}$$

This shows explicitly how the track density depends on the angle ψ between the plane of observation and a prismatic face of the crystal.

When the length distribution is isotropic, we may put $\mu_1 = \mu_2 = 0$ and $\mu_0 = \mu$ to get

$$\rho = \tfrac{1}{2}\tau\mu$$

in agreement with (2.12), and confirming that, in the isotropic case, the track density is the same for any plane.

7.8.2 Equivalent isotropic length

Using (7.41) and (2.13), the equivalent isotropic length is

$$\mu^* = \mu_0 + \tfrac{4}{3\pi}\left(\cos\psi + \psi\sin\psi\right)\mu_1 + \tfrac{1}{4}\left(1 + \sin^2\psi\right)\mu_2. \tag{7.42}$$

When the plane of observation is a prismatic face (i.e., $\psi = 0$), this becomes

$$\mu^* = \mu_0 + \tfrac{4}{3\pi}\mu_1 + \tfrac{1}{4}\mu_2, \tag{7.43}$$

which can also be derived directly by substituting (7.37) into (2.17). Comparing this formula with the true mean length μ given by (7.40) confirms that μ^* is less than or equal to μ.

7.8.3 Horizontal confined tracks

Substituting (7.37) into the first formula in (7.36) leads to

$$\mu_{hc} = E[X_{hc}] = \mu_0 + \tfrac{2}{\pi}\mu_1 + \tfrac{1}{2}\mu_2$$

for the mean length of horizontal tracks, which differs slightly from the overall mean length (7.40).

7.8.4 Semi-track lengths

Substituting (7.38) into equation (7.33) and evaluating the integrals as above give the following formula for the mean semi-track length:

$$\begin{aligned}
\mu_s &= \frac{1}{\pi\mu^*}\sum_{j=0}^{k}\gamma_j\, B\left(\tfrac{3}{2}, \tfrac{j+1}{2}\right)\\
&= \frac{1}{2\mu^*}\left(\gamma_0 + \tfrac{4}{3\pi}\gamma_1 + \tfrac{1}{4}\gamma_2 + \cdots\right).
\end{aligned} \tag{7.44}$$

In the isotropic case, only the leading term is non-zero, where $\gamma_0 = E(l^2) = \mu^2 + \sigma^2$ and $\mu^* = \mu$ so μ_s reduces to (7.10) as before.

7.8.5 Projected semi-track lengths

Similarly, substituting (7.38) into equation (7.23) it can be shown that the mean projected semi-track length reduces to

$$\begin{aligned}
\mu_{ps} &= \frac{1}{\pi\mu^*}\sum_{j=0}^{k}\gamma_j\,\tfrac{j+2}{j+3}\, B\left(\tfrac{3}{2}, \tfrac{j+1}{2}\right)\\
&= \frac{1}{3\mu^*}\left(\gamma_0 + \tfrac{3}{2\pi}\gamma_1 + \tfrac{3}{10}\gamma_2 + \cdots\right)
\end{aligned} \tag{7.45}$$

after integrating the term containing $\log(\sec\phi + \tan\phi)$ by parts.

7.8.6 Projected angle to the c-axis

For the probability density function of ω, the angle between the projected track and the c-axis, substituting $E[l|\phi]$ from (7.37) into (7.28), gives

$$f_{ps}(\omega) = \frac{4}{\pi\mu^*}\left(\tfrac{1}{2}\mu_0 + \tfrac{1}{3}\mu_1\cos\omega + \tfrac{1}{4}\mu_2\cos^2\omega\right).$$

This still depends on ω, though not as strongly as $E[l|\phi]$ depends on ϕ. Nevertheless, strong anisotropy of the track length distribution should be evident, not only in the distribution of r but also in that of ω.

7.8.7 Projected length at a given angle

Finally, the formulae for the conditional moments of r given ω given by (7.29) and (7.30) can also give explicit expressions using the fact that

$$\int_\omega^{\frac{\pi}{2}} \sin\phi\cos^j\phi\, d\phi = \frac{\cos^{j+1}\omega}{j+1}.$$

From (7.38) and (7.39) these lead to

$$E(X_{ps}|\omega) = \frac{1}{2}\sum_{j=0}^{k} \frac{\gamma_j\cos^j\omega}{j+3} \bigg/ \sum_{j=0}^{2} \frac{\mu_j\cos^j\omega}{j+2}$$

and

$$E(X_{ps}^2|\omega) = \frac{1}{3}\sum_{j=0}^{k} \frac{\delta_j\cos^j\omega}{j+4} \bigg/ \sum_{j=0}^{2} \frac{\mu_j\cos^j\omega}{j+2}.$$

7.9 A two-component mixture of anisotropic lengths

In some applications it is useful to consider mixtures of two distributions of lengths and angles. For example, this would apply to a thermal history such as that in Figure 1.3, panel (d), in Chapter 1.

Imagine two groups of uranium atoms: a proportion p that produced tracks prior to some time t_u which have been partially annealed and then cooled, and a proportion $1-p$ that produced tracks after t_u which have not been so heavily annealed. Let $f_{\phi 1}(l)$ and $f_{\phi 0}(l)$ be the distributions of l, for tracks at angle ϕ, for the two groups. Then $f_\phi(l)$ is a two-component mixture

$$f_\phi(l) = (1-p)f_{\phi 0}(l) + pf_{\phi 1}(l). \tag{7.46}$$

The lengths l will vary within each group, but tracks from the earlier group will tend to be shorter, and more so at higher angles ϕ. From equation (2.8), the parameter p satisfies

$$1-p = \frac{p(t_u)}{p(t_0)} = \frac{1-e^{-\lambda t_u}}{1-e^{-\lambda t_0}} \approx \frac{t_u}{t_0},$$

where λ is the total ^{238}U decay rate and t_0 is the total time over which tracks have been forming. So estimating p gives information about t_u/t_0. Also $f_{\phi 1}(l)$ contains information about the maximum temperature experienced by tracks formed before t_u. This can be regarded as a more realistic version of the two-length model in Section 7.2.

What is the joint probability distribution of the projected semi-track lengths and angles? Let $f_{ps0}(r,\omega)$ and $f_{ps1}(r,\omega)$, respectively, denote the probability density functions of r and ω arising from each group of uranium atoms. These are of the form (7.27) with $f_\phi(l)$ given by $f_{\phi 0}(l)$ or $f_{\phi 1}(l)$, respectively. Substituting (7.46) into (7.27) shows that the overall joint distribution of r and ω is also a two-component mixture with probability density

$$f_{ps}(r,\omega) = (1-q)f_{ps0}(r,\omega) + qf_{ps1}(r,\omega) \qquad (7.47)$$

but with mixing proportions $1-q$ and q, where

$$q = \frac{p\mu_1^*}{(1-p)\mu_0^* + p\mu_1^*} = \frac{p\mu_1^*}{\mu^*}. \qquad (7.48)$$

Here μ^* is the equivalent isotropic length, given by (2.17), and μ_0^* and μ_1^* are similarly defined for each component, so that

$$\mu^* = (1-p)\mu_0^* + p\mu_1^*.$$

So for tracks intersecting a prismatic face, the component distributions are mixed in *length-biased* proportions $1-q$ and q, defined in terms of equivalent isotropic lengths.

The same result holds for the marginal distributions of r and ω, for the conditional distribution of r given ω, and for other related distributions. For example, probability density function of semi-track lengths t is

$$f_s(t) = (1-q)f_{s0}(t) + qf_{s1}(t), \qquad (7.49)$$

where q is given by (7.48).

7.10 Quantitative effects of anisotropy

Some quantitative effects of anisotropy can be illustrated using calculations taken from Galbraith and Laslett (1988). These used $E[l|\phi]$ given by (7.37) with the coefficients μ_0, μ_1 and μ_2 in Table 7.1. These coefficients were obtained by fitting equation (7.37) to measurements of lengths and angles to the c-axis of horizontal confined tracks for a number of samples of Durango apatite that were heated for 1 hour at various temperatures. The tracks measured by Paul Green were tints and those by Andrew Gleadow were tincles. All were etched for a slightly longer time than is usual nowadays in routine work (in which both track lengths and densities are measured).

To calculate the means and standard deviations of the different types of track length measurement from the above formulae, one needs, in addition to $E[l|\phi]$, formulae for $E[l^2|\phi]$ and $E[l^3|\phi]$. Such might be obtained by fitting curves to the observed graphs of l^2 and l^3 against ϕ. However, the calculations shown here used a simpler method based on the assumptions that $\mathrm{var}[l|\phi]$ does

Table 7.1 *Coefficients μ_0, μ_1 and μ_2 from Galbraith and Laslett (1988). These were obtained by fitting (7.37) to measurements of l and ϕ from annealing experiments by Paul Green (Green et al., 1986) and Andrew Gleadow (Laslett et al., 1984). Annealed samples were heated for 1 hour at the temperatures shown.*

	Green (tints)				Gleadow (tincles)		
	μ_0	μ_1	μ_2		μ_0	μ_1	μ_2
Unannealed	16.72	−2.81	3.55	Unannealed	16.18	0.21	0.55
260°C	14.70	−1.62	2.62	300°C	13.28	−2.72	4.29
310°C	12.45	−0.42	2.15	325°C	11.28	1.43	0.63
336°C	9.94	−2.80	5.13	350°C	−0.07	18.78	−8.47
352°C	1.11	12.25	−2.23				

not depend on ϕ and that f_ϕ is symmetric. Then, writing $\mathrm{var}[l|\phi] = \sigma_0^2$, it can be shown that

$$E[l^2|\phi] = \sigma_0^2 + \left(E[l|\phi]\right)^2 \quad \text{and} \quad E[l^3|\phi] = 3\sigma_0^2 E[l|\phi] + \left(E[l|\phi]\right)^3. \quad (7.50)$$

So $E[l|\phi]$, $E[l^2|\phi]$ and $E[l^3|\phi]$ are all polynomials in $\cos\phi$ with coefficients depending on μ_0, μ_1, μ_2 and σ_0^2. Hence all of the necessary integrals may be evaluated explicitly in terms of just these four parameters.

In fact, for the present purpose it is not necessary to evaluate all of these integrals explicitly, because they may be calculated empirically as weighted averages over ϕ of the various conditional moments of l given ϕ. Furthermore it is not necessary to use polynomials in $\cos\phi$ — any smooth curves that behave in the right way will do. The essential points are that the curve should extend over the full range of ϕ and that the correct weights are used.

Table 7.2 shows the equivalent isotropic length for tracks at each of five annealing temperatures, along with the theoretical mean, standard deviation and coefficient of variation of horizontal confined track lengths, semi-track lengths and projected semi-track lengths. These were calculated from equations (7.19) for μ^*, (7.36) for horizontal confined tracks, (7.33) and (7.34) for semi-tracks and (7.23) and (7.24) for projected semi-tracks. All of these formulae were evaluated using (7.37) and (7.50) with the coefficients for Green's data in Table 7.1 and $\sigma_0 = 0.84$ microns.

7.10.1 Means and standard deviations of track lengths

The first part of Table 7.2 gives theoretical means, standard deviations and coefficients of variation for horizontal confined tracks. The empirical distribution of unannealed or lightly annealed confined track lengths in apatite is very tight, with a mean of about 16 microns and a standard deviation of less than 1 micron (coefficient of variation $\approx 7\%$). But for tracks annealed to lengths below about 10.5 microns, the anisotropic nature of annealing increases the dispersion considerably. These features are reflected in the theoretical values

Table 7.2 *Track length parameters calculated from (7.23), (7.24), (7.33), (7.34) and (7.36) using (7.37) and (7.50) with $\sigma_0 = 0.84$ and the coefficients for Green's data in Table 7.1. From Galbraith and Laslett (1988).*

Horizontal confined tracks, from (7.36)

	μ^* microns	mean microns	s.d. microns	c.v. %
Unannealed	16.41	16.71	0.95	6
260°C	14.67	14.98	0.95	6
310°C	12.81	13.26	1.05	8
336°C	10.03	10.72	1.30	12
352°C	5.75	7.79	3.12	40

Semi-tracks, from (7.33) and (7.34)

	$\frac{1}{2}\mu^*$	mean	s.d.	c.v.
Unannealed	8.21	8.23	4.78	58
260°C	7.33	7.36	4.28	58
310°C	6.40	6.44	3.76	58
336°C	5.02	5.07	2.99	59
352°C	2.88	3.57	2.52	71

Projected semi-tracks, from (7.23) and (7.24)

	$\frac{1}{3}\mu^*$	mean	s.d.	c.v.
Unannealed	5.47	5.51	3.93	71
260°C	4.89	4.94	3.53	72
310°C	4.27	4.35	3.13	72
336°C	3.34	3.46	2.54	73
352°C	1.92	2.76	2.29	83

in Table 7.2, where the coefficient of variation rises to 40% as the mean length reduces to 7.8 microns.

Figure 7.6, taken from Galbraith and Laslett (1988), is a plot of standard deviation against mean confined track length, showing the theoretical values from Table 7.2 for Green's data (solid squares) and corresponding theoretical values for Gleadow's data (open squares). Also plotted are some observed means and standard deviations (open circles) obtained from a large number of other experiments in which angles to the c-axis were not measured. The agreement is very good, in spite of the various observational biases mentioned earlier. In particular, the observed empirical relation between the standard deviation and mean length can be completely explained by the anisotropy of $E[l|\phi]$.

Table 7.2 also gives the corresponding quantities for semi-track lengths and for projected semi-track lengths, obtained using (7.33), (7.34), (7.23) and

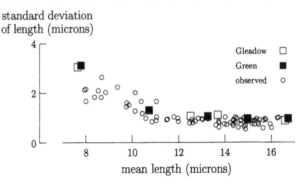

Figure 7.6 *Relation between standard deviation and mean length in Table 7.2 for horizontal confined tracks (from Galbraith and Laslett, 1988). The circles correspond to observed means and standard deviations from Green et al. (1986, Figure 3) that were obtained from an extensive study with Durango apatite, in which angles to the c-axis were not measured.*

(7.24). The mean semi-track length is about half the mean confined track length, and the coefficient of variation is substantially greater. For unannealed semi-tracks the coefficient of variation is 58%, which is very close to the theoretical figure obtained earlier for the "ideal" rectangular distribution that would obtain if all tracks had the same length. For projected semi-tracks, which have a mean length about one third that of confined tracks, the coefficient of variation is greater still. Again, for the unannealed tracks, the coefficient of variation of 71% is close to that for a triangular distribution. These calculations emphasise again that the distribution of confined track lengths is much more informative as an indicator of thermal history than is the distribution of semi-track lengths or projected semi-track lengths.

7.10.2 Relation between mean track length and track density

Figure 7.7, also from Galbraith and Laslett (1988), shows the theoretical mean track lengths in Table 7.2 plotted against the equivalent isotropic length μ^* given by (7.19) where μ^* is proportional to ρ given by (2.16).

This shows the departure from the proportionality relationship (2.12) due to anisotropic annealing, where track density decreases faster than mean length as the degree of annealing increases. This departure is substantial for μ^* between 5 and 10 microns, whether confined tracks, semi-tracks or projected semi-tracks are measured. Also there is excellent agreement in the relationship using the coefficients from the two different sets of data — one using tints and the other tincles.

This type of departure from (2.12) has been observed empirically from direct measurements of lengths and densities (e.g., Watt and Durrani, 1985;

mean length in microns

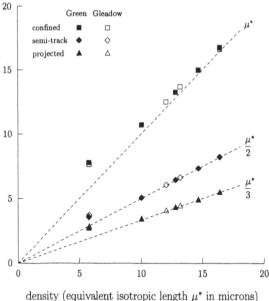

density (equivalent isotropic length μ^* in microns)

Figure 7.7 *Relation between mean fission track length and track density on a prismatic face of an apatite crystal (from Galbraith and Laslett, 1988). Mean lengths in Table 7.2 are plotted against the equivalent isotropic length μ^*. The dashed lines indicate the proportionality relation (2.12).*

Green, 1988 and Donelick *et al.*, 1999). Such data will reflect variation due to experimental sources in measuring track lengths (including etching variation and sampling biases discussed in Chapter 8) and in measuring track densities (see Chapter 3) in addition to the inherent variation described by the Poisson line segment model. Figure 7.7 shows the extent to which the departure from proportionality is inherent in the theoretical distributions derived in this chapter, and provides a baseline for assessing these other sources of variation.

Figure 7.7 also illustrates a useful methodological point. In Chapter 8 we will see that, because of sampling biases, the distributions of sampled orientations of tints and tincles are different, leading (among other things) to different *empirical* length-density relationships. But in Figure 7.7, the open and closed squares lie on practically the same curve — i.e., the length-density relation is the same for tints and tincles. This has been achieved by measuring the orientation ϕ and using $E[l|\phi]$ with the same weights for both — in other words, by standardising the length measurements to the same orientation distribution. This principle is useful more generally.

7.11 Parametric models for length against angle

The function $E[l|\phi]$ arises so often that it would be very useful to have a parametric formula for it, and preferably one that has only a small number of unknown parameters. It is fairly clear that this function should decrease as ϕ increases, but there does not appear to be any hard theory to suggest what form it should take.

A simple form is

$$E[l|\phi] = \eta + \xi \cos \phi, \qquad (7.51)$$

which has the advantage of mathematical tractability. Practically all formulae involving $E[l|\phi]$ can be evaluated explicitly, as in Section 7.8, and can therefore easily be used in simulations and other studies. This equation may be re-parameterised as

$$E[l|\phi] = \mu_a + (\mu_c - \mu_a) \cos \phi, \qquad (7.52)$$

where μ_c and μ_a are the mean lengths of tracks parallel and perpendicular to the c-axis, respectively. Thus μ_c and μ_a are defined by

$$E[l|\phi = 0] = \mu_c \quad \text{and} \quad E[l|\phi = \tfrac{\pi}{2}] = \mu_a$$

and it is assumed that $\mu_c \geq \mu_a$. In the isotropic case, $\mu_a = \mu_c$ of course. From Section 7.8.1, the mean track length over all orientations is given by

$$\mu = \eta + \tfrac{1}{2}\xi = \tfrac{1}{2}(\mu_c + \mu_a)$$

and from Section 7.8.2 the equivalent isotropic length is

$$\mu^* = \eta + \tfrac{4}{3\pi}\xi = \tfrac{4}{3\pi}\mu_c + (1 - \tfrac{4}{3\pi})\mu_a ,$$

which is a weighted average of μ_c and μ_a. Hence μ is the value of $E[l|\phi]$ when $\cos \phi$ equals $\tfrac{1}{2}$, i.e., when $\phi = \tfrac{\pi}{3}$, or 60 degrees. Similarly, μ^* is the value of $E[l|\phi]$ when $\cos \phi$ equals $\tfrac{4}{3\pi}$, which corresponds to an angle ϕ of about 65 degrees.

Equation (7.51) was used by Galbraith *et al.* (1990) and Laslett *et al.* (1994) where, in addition, η and ξ were specified in terms of the mean length μ by

$$\eta = -5.18 + 1.42\mu - \sqrt{11.48 - 2.13\mu + 0.10\mu^2} \qquad \text{and}$$
$$\xi = 2(\mu - \eta) . \qquad (7.53)$$

The second of these equations ensures that $\mu = \eta + \tfrac{1}{2}\xi$ as required. The first was fitted to data from a series of experiments using Durango apatite, taking into account some observational effects discussed in Chapter 8. Details of this exercise are noted in Laslett (1993). The distributions in Figure 7.5 were calculated using this model.

Donelick (1991) suggested a different model, where

$$E[l|\phi] = \left(\frac{\sin^2\phi}{\mu_a^2} + \frac{\cos^2\phi}{\mu_c^2} \right)^{-\frac{1}{2}} . \qquad (7.54)$$

If plotted in polar coordinates (l, ϕ), this is the equation of an ellipse centered

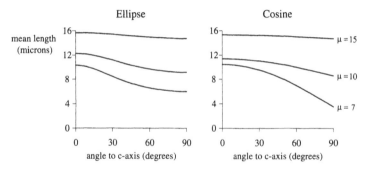

Figure 7.8 *Elliptic and cosine forms for* $E[l|\phi]$.

at the origin with its major axis parallel to the x-axis (here the crystallo-graphic c-axis). In polar coordinates, increasing anisotropy is represented as increasing departure from a circle. This model has the merit that it is defined quite naturally on the whole circle $0 \le \phi \le 2\pi$, though it is only the quadrant $0 \le \phi \le \frac{\pi}{2}$ that is relevant in practice. A disadvantage is that most of the formulae in which $E[l|\phi]$ appears involve elliptic integrals and cannot be evaluated explicitly. An exception is the mean length, given in equation (7.56) below, though even there the formula is not simple.

Equation (7.54) may be written in the alternative form

$$E[l|\phi] = \left(\frac{1}{\mu_a^2} - \left(\frac{1}{\mu_a^2} - \frac{1}{\mu_c^2} \right) \cos^2\phi \right)^{-\frac{1}{2}}, \qquad (7.55)$$

which is more convenient for fitting to empirical measurements. From this, an explicit formula for the mean track length μ, when $\mu_c > \mu_a$, can be found using an integral tabulated in Gradshteyn and Ryzhik (1980, page 387, §3.676,1). Hence

$$\mu = \frac{\mu_c \mu_a}{\sqrt{\mu_c^2 - \mu_a^2}} \arctan \sqrt{\frac{\mu_c^2 - \mu_a^2}{\mu_a^2}}. \qquad (7.56)$$

The equivalent isotropic length μ^* cannot be obtained explicitly, though. Approximate formulae can be obtained, if desired, by expanding (7.55) in a power series in $\cos^2\phi$ and integrating term by term.

Figure 7.8 shows example curves given by (7.55) and (7.52) corresponding to underlying mean lengths μ of 15, 10 and 7 microns. The two forms do not differ greatly numerically, especially at higher mean lengths, but there is a qualitative difference in their behaviour at high angles. In both cases, $E[l|\phi]$ decreases as ϕ increases, as one would want, but the slope gets flatter for the ellipse and steeper for the cosine. In general, empirical studies have suggested that the *rate* of annealing increases with the amount of annealing, which would argue for the latter behaviour. Empirical measurements of l and ϕ do not seem consistently to favour one over the other, though. This may be partly because, for high mean lengths the two forms do not differ much and for low mean lengths other observational factors come into play (see for example Figure 8.1 in the next chapter).

A remark about graphical representation may be useful here, as the reader may wonder why (7.54) has not been plotted in polar coordinates. While this is the natural representation of an ellipse, it is a poor one for seeing how the mean length depends on angle. In particular, the difference in behaviour clearly seen in Figure 7.8 would be scarcely visible. Likewise, plotting observed lengths and angles in polar coordinates makes it hard to assess how well a given curve agrees with the data.

For the purposes of fitting equations and for calculating higher moments, it is also useful to have a simple parametric model for the variance of lengths at a given angle. The simplest model is to assume that this variance is the same for all angles, i.e., $\text{var}[l|\phi] = \sigma_0^2$, say. A more general formula is

$$\text{var}[l|\phi] = \sigma_0^2 + \sigma_1^2 \left(1 - \text{E}[l|\phi]/\mu_0\right)^\kappa \tag{7.57}$$

for suitable μ_0, σ_0, σ_1 and κ. Here μ_0 and σ_0 represent the theoretical mean and standard deviation of unannealed track lengths, values of which may be obtainable from laboratory data. Experience suggests that suitable values of κ are 1 or 2. In either case, $\text{var}[l|\phi] = \sigma_0^2$ for unannealed tracks and increases both with increasing annealing and with increasing angle. Here σ_1 would need to be estimated also.

7.12 Bibliographic notes

This chapter uses material from Galbraith and Laslett (1988), Laslett, Galbraith and Green (1994) and Laslett and Galbraith (1996a).

The distributions of semi-track length and projected semi-track lengths for the single-length case were derived by Dakowski (1978) and those for the general isotropic case by Laslett et al. (1982), who also introduced the concept of length-biased sampling in this context. The relation between mean length and track density on a prismatic face for the anisotropic case was given by Laslett et al. (1984). Derivations of the anisotropic distributions of X_s and X_{ps} were given in Galbraith and Laslett (1988).

CHAPTER 8

Observational features of track measurements

Earlier chapters showed how counts of numbers of fission tracks may provide information about the age of a host rock or the timing of geological thermal events. In addition, measurements of lengths and orientations of tracks may provide information about the *temperatures* experienced by a host rock, as noted in Chapter 1. In theory, the statistical distribution of track lengths may be related to the whole thermal history experienced by the sample — though the nature of this relationship is such that it is not feasible to reconstruct the full detail, but rather to estimate some key parameters.

The process of revealing fission tracks by chemical etching of a crystal surface does not produce a proper random sample from the length distribution. Inferring thermal history information from track measurements thus involves two conceptual steps. The first is to relate measurements made on a sample to the true length distribution $f(l)$, and the second is to relate $f(l)$ to the thermal history. The second step is discussed briefly in Chapter 9. With respect to the first step, Chapter 7 dealt with theoretical properties, while this chapter looks at observational features. What measurements are easily available and what do they tell us about the true distribution of track lengths and orientations within a crystal? As before, the line segment model provides a framework for answering such questions.

8.1 Horizontal confined tracks

8.1.1 Tints and tincles

The most direct information about $f(l)$ can be obtained by measuring horizontal confined tracks. These are confined tracks that are parallel to the plane of observation, which we take to be a prismatic face of an apatite crystal. A confined track may be revealed by etching because it intersects a semi-track or fracture down which the etchant has penetrated. Examples are shown in Figure 1.2 in Chapter 1. A confined track that is etched because it intersects a semi-track is called a *tint* (track-in-track), while one revealed by intersecting a fracture is called a *tincle* (track-in-cleavage).

Tracks are seen through a microscope linked to a computer. Once a horizontal confined track is identified, the location of each of its ends may be recorded by placing a mouse spot there and clicking a button. This allows the length of the track to be calculated, as projected onto the plane of observation. For a

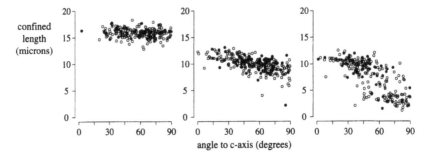

Figure 8.1 *Scatter plots of measured lengths and angles to the c-axis for three samples of induced horizontal confined tracks in Durango apatite. Left panel, unannealed; middle panel, heated for 1 hour at 300° C; right panel, heated for 1 hour at 343° C. Open circles denote tints and filled circles denote tincles.*

Table 8.1 *Summary statistics for the data in Figure 8.1. Number of tracks n, mean length $\bar{y}$ and standard deviation of length s_y in microns.*

		n	$\bar{y}$	s_y
Unannealed	Tincles	27	16.14	0.80
	Tints	173	15.79	0.88
Annealed, 300° C	Tincles	57	10.13	1.63
	Tints	203	9.97	1.42
Annealed, 343° C	Tincles	58	8.27	3.08
	Tints	166	7.36	3.12

horizontal track this equals the full length. The locations of two points along the crystallographic c-axis may be recorded in the same way, which allows the angle between the track and the c-axis to be calculated.

Figure 8.1 shows scatter plots of lengths and angles to the c-axis for three samples of horizontal confined tracks. Tints and tincles are distinguished by different plotting symbols. Table 8.1 gives the number of tracks measured along with the mean and standard deviation of the lengths, separately for tints and tincles, for each sample. As the amount of annealing increases, the mean length decreases and the standard deviation increases. The empirical distributions of tints and tincles look similar in Figure 8.1, but tincles tend to be longer on average than tints in the same sample.

8.1.2 Empirical distributions of lengths and angles

Figure 8.2 shows the scatter plots for the tints only from Figure 8.1 along with histograms of their lengths and angles to the c-axis. These illustrate the main effects of thermal annealing on fission track lengths as well as some

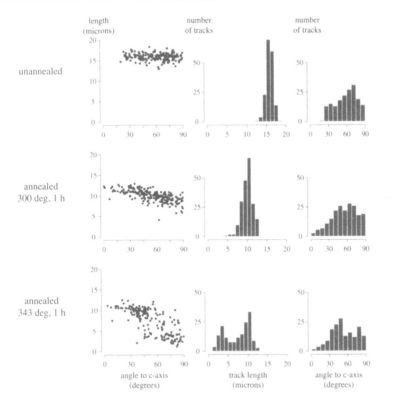

Figure 8.2 *Scatter plots and histograms of measured lengths and angles to the c-axis for three samples of induced horizontal confined tints in Durango apatite.*

observational effects. For the unannealed sample the measured lengths vary with a small standard deviation (less than 1 μm) about a high mean (about 16 μm), independently of orientation. Their length distribution is approximately isotropic. For the annealed samples the lengths are shorter and more dispersed. Also the length distribution is anisotropic: tracks at a higher angle to the c-axis tend to be shorter than those at a lower angle. Both of these effects increase with the amount of annealing. In addition, there are some very short measured lengths in the highly annealed samples. Here the length distribution may have two components: very short "gapped" track fragments and non-gapped tracks whose lengths vary around a smooth anisotropic curve. Furthermore, the orientations are not uniformly distributed over the range 0 to 90 degrees, and there are relatively few tints at low angles to the c-axis.

8.1.3 Criteria for horizontal tracks

In common practice two simple criteria have been used to identify whether a confined track is acceptably close to horizontal:

(a) tracks giving a bright reflection: etched tracks showing a bright image in reflected light are at a low angle to the horizontal, and their projected length onto the observation plane is very close to their full length. For tracks within 15° to the horizontal, the relative difference is less than 4%.

(b) tracks in full focus: etched tracks fully in focus have their projected length very close to their full length. With a focal depth of 1 micron, a track whose projected length is only 3 microns has a full length between 3 and 3.16 microns, a maximum relative error of only 5%.

Criterion (a) means that the observer effectively accepts all tracks whose dip angle ϑ to the horizontal is less than some fixed ϑ_0, although the intensity of reflection may depend on other factors, such as the thickness of the etched track. Criterion (b) means that a track of length l at angle ϑ to the horizontal is accepted if $l \sin \vartheta \leq \delta$, where δ is the depth of focus. Thus $\vartheta \leq \arcsin (\delta/l)$, which is not a fixed upper bound. So the criteria (a) and (b) are different and will give rise to samples with different statistical properties.

When $\delta = 1$ micron and $\vartheta_0 = 15°$, criterion (a) includes (b) for tracks longer than about 3.5 microns. Hence (a) gives rise to larger sample sizes and is the method recommended by Laslett *et al.* (1982); see also Galbraith and Laslett (1988). For tracks of length 15 microns, the use of definition (a) with $\vartheta_0 = 15°$ compared with (b) with $\delta = 1$ micron increases the sample size of included tracks by a factor of 4. In real samples, the observer typically wishes to maximise sample size, so definition (b) alone is not sufficient. For example it would seem foolish to omit a highly informative 8 micron track at 10° to the horizontal simply because it is slightly out of focus.

8.1.4 Gapped tracks

In the bottom left panel of Figure 8.2 there some very short length measurements (less than 5 μm). Most of these are at high angles to the c-axis but there are some at angles below 45°. This is a common feature of confined length measurements in highly annealed samples. Then the histogram of lengths often has a long lower tail or a second mode below 5 μm, as can be seen in the bottom middle panel of Figure 8.2. These features indicate that a sample has been heavily annealed. But they do not necessarily imply, for example, that it has experienced more than one thermal event: in the present case, all tracks have experienced the same amount of heat.

Some of these short lengths, particularly those at lower angles, may be due to a phenomenon called "gapping" — track fragmentation (shortly before there is complete repair) along with some associated etching effects. Green *et al.* (1986) showed that some tracks that had been etched to 3–4 μm would increase to 7–8 μm with a small amount of further etching. At the same time, more short tracks may appear, so it is necessary to re-measure exactly the same tracks after the second etching in order to see this.

From a statistical point of view it is not clear how best to use these very short length measurements. Possible approaches include:

(a) regarding the distribution of lengths at a given angle, as having two components (gapped or non-gapped), or

(b) treating all length measurements at face value but with variance increasing with angle and with the amount of annealing.

The choice may depend on the particular circumstances or purpose. For example, when using such data to estimate $E[l|\phi]$ one needs to be clear that one is modelling just the effect of annealing and not the effect of etching also.

In this context, Galbraith *et al.* (1990) and Laslett *et al.* (1994) used the approach (a) with $E[l|\phi] = \eta + \xi \cos \phi$ (see Section 7.11) where this was taken to apply to fully etched (non-gapped) tracks only. These studies were concerned with properties of semi-tracks, which are etched from the crystal's surface and therefore will take a shorter time to become fully etched than will the confined tracks to which the model for $E[l|\phi]$ is fitted. This equation was further reduced to depend on a single parameter, the mean length μ, according to (7.53). With respect to the approach (b) equation (7.57) gives a useful form for the variance of lengths of tracks at a given angle.

8.1.5 Features of the empirical length distribution

Suppose that an observer accepts all tracks within a given dip angle to the horizontal and that the measured length is the full etchable length. Then the empirical distribution of horizontal confined tracks has a number of general features.

Firstly, it is subject to *length-biased* sampling, where longer tracks are over-represented (Laslett *et al.*, 1982). Secondly, there is an *orientation bias* that favours the sampling of tints at a high angle to the c-axis. This is partly a consequence of the etching process (Galbraith *et al.*, 1990; Donelick *et al.*, 1999). In annealed samples, such tracks tend to be shorter, so this orientation bias acts in the opposite direction to length bias. For tincles, angles of host fractures are not uniform, and their thickness also depends on their orientation because of the anisotropic etching process.

A third feature is that the requirement that both ends of a tincle are visible either side of a fracture biases the sample. This is called *fracture-thickness bias* (Laslett *et al.*, 1982, Galbraith *et al.*, 1990) and has the greatest effect for very short tracks. Fourthly, tincles are on average longer than tints in the same sample. This would be expected from orientation bias and fracture-thickness bias, but there may also be other reasons. Fifthly, short tracks very near the surface are more likely to be totally confined than longer tracks at the same dip angle (Laslett *et al.*, 1982). Similarly, shorter tracks near a grain edge are more likely than longer tracks to be totally confined within the grain.

Some of these sampling features are quite subtle, but they are well understood and have only small effects if all lengths are greater than about 8 microns. We consider them in more detail below.

8.2 Length bias

Length-biased sampling of confined tracks arises naturally because for a confined track to be etched and hence observed, it must intersect a fracture or
a semi-track. A uranium atom associated with a longer track can be further
away from a fracture or host semi-track and still be etched, so that longer
tracks will be over-represented in the etched sample. In theory the sampling
bias factor is proportional to the volume of the region in which the etched
track's associated uranium atom can lie, and hence, for a horizontal track, is
proportional to the track's length. Figure 8.3 illustrates this. Here uranium
atoms are distributed uniformly either side of a fracture (of negligible width)
and produce equal numbers of horizontal confined tracks of three different
lengths (15, 10 and 5 microns) all at the same angle to a host fracture. But
the relative numbers that intersect the fracture, and are therefore revealed,
are 3:2:1, in proportion to the three lengths.

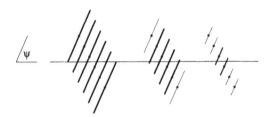

Figure 8.3 *Length-biased sampling of horizontal tincles. The horizontal line repre-*
sents the intersection of a thin fracture in the crystal with the observation plane.
Line segments represent fission tracks of three different lengths parallel to the plane
of observation and at the same angle ψ to the fracture. Bold line segments are those
that intersect the fracture and will thus be etched.

More generally, for the isotropic line segment model, where latent track
lengths l have probability density function $f(l)$ and mean length μ, the distribution of lengths of horizontal tincles or tints has probability density $g(l)$
given by

$$g(l) = lf(l)/\mu, \quad 0 < l < \infty. \tag{8.1}$$

This is called the *length-biased density* of l — the probability density that
results when lengths from $f(l)$ are sampled with probability proportional
to length. For non-horizontal tracks the sampling bias is more complicated
(Laslett *et al.*, 1982).

To get a feel for the practical consequences of equation (8.1), Figure 8.4
shows three probability density functions $f(l)$ and their corresponding length-
biased densities $g(l)$. In case (a), $f(l)$ is Normal with mean 15 and standard
deviation 1 μm, and $g(l)$ is practically identical, with mean 15.067 and standard deviation 0.998 μm. In case (b), $f(l)$ is normal with mean 10 and standard

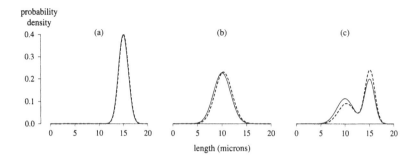

Figure 8.4 *Length-biased densities g(l) (dashed lines) for three hypothetical isotropic length distributions f(l) (solid lines): (a) Normal with mean μ = 15 and standard deviation σ = 1; (b) Normal with μ = 10 and σ = 1.75; and (c) a 50:50 mixture of (a) and (b).*

deviation 1.75 μm, and $g(l)$ does differ slightly, with mean 10.306 and standard deviation 1.723 μm. In case (c), $f(l)$ is a 50:50 mixture of the two normal distributions in (a) and (b). The length-biased density $g(l)$ is a mixture of the two corresponding length-biased densities, but in the ratio 40:60, rather than 50:50. That is, the proportions of tracks in each component are $10/(10+15)$ = 0.40 and $15/(10+15)$ = 0.60, in proportion to the two mean lengths. In general, length-biased sampling has very little effect on a distribution with low dispersion, but can have a substantial effect on a distribution with high dispersion.

For the anisotropic line segment model the same principle applies to horizontal tracks that are all at the same angle ϕ to the c-axis. For such tracks the length-biased density is

$$g_\phi(l) \;=\; lf_\phi(l)/\mu(\phi), \quad 0 < l < \infty, \tag{8.2}$$

where $\mu(\phi) = E[l|\phi]$ is the mean length of tracks at angle ϕ. Integrating $g_\phi(l)$ over the distribution of ϕ does not lead to (8.1) in general because $\mu(\phi)$ varies with ϕ.

8.3 The Loaded Dog experiments

It is convenient to describe here some experiments carried out by Paul Green, called the Loaded Dog experiments because they were designed in a North Melbourne pub of that name. They were reported briefly in Galbraith *et al.* (1990) and in Laslett *et al.* (1994). We will use them to illustrate some of the concepts in this chapter.

Induced fission tracks were created in ten samples of Durango apatite, five using a high neutron fluence and five using a low neutron fluence, and a sample of each was set aside as unannealed controls. The remaining eight samples were thermally annealed for 1 hour at one of two temperatures (300°C or 343°C) and four annealed control samples were set aside, one for each combination of neutron fluence and temperature. Using the higher neutron fluence produced larger numbers of induced tracks, while using the higher temperature produced shorter tracks.

In the remaining four samples, further induced fission tracks were created, by irradiating them using a yet lower fluence, to produce mixtures of annealed and unannealed tracks within each crystal grain. This produced four two-component mixture populations of long and short tracks all with the same mean length μ_0 for the long component, where $\mu_0 \approx 16$ μm. For the two mixtures in which tracks were annealed at 300°C, the short tracks had a mean length of $\mu_1 \approx 10$ μm. But they had different mixing proportions $p = 15/16$ or $p = 5/6$, depending on whether the original tracks were induced with the high or a low fluence. Here p denotes the proportion of short (annealed) tracks in the population. For the other two mixtures (in which tracks were annealed at 343°C) the short tracks had a mean length μ_1 of about 8 μm. Again, in one of these the proportion of short tracks was $p = 15/16$ and in the other, $p = 5/6$. Samples from each component population were also available from the unannealed and annealed controls.

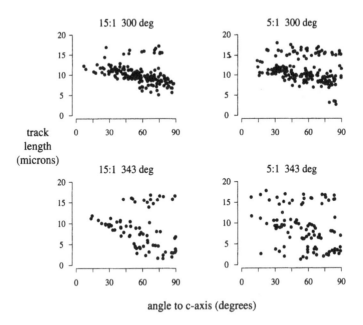

Figure 8.5 *Scatter plots of track lengths against angles to the c-axis for horizontal confined tracks (tints) in four two-component mixture populations in the Loaded Dog experiments.*

Horizontal confined tracks were identified and measured using standard laboratory practice. Figure 8.1, referred to earlier, shows the scatter plots of the lengths against angle for the unannealed and annealed control samples, where each two samples generated with different fluences have been combined, and Table 8.1 gives some summary statistics, while Figure 8.2 shows graphs of the measurements on all *tints* in these samples. Figure 8.5 here shows track length plotted against angle ϕ for the tints in each of the two-component mixtures. It is easy to identify observations from the separate components visually and to compare them with those in Figure 8.2.

We discuss these data further below. Also, about 500 projected semi-track lengths and angles were measured in some samples, which are discussed in Section 8.9.

8.4 Empirical verification of length bias

Formulae (8.1) and (8.2) are based on the idealised isotropic and anisotropic line segment models, so it is of some interest to see how they apply in practice. It is not straightforward to do this because the true length distribution $f(l)$ is unknown and because there are other observational effects to be accounted for. We look at two methods from Galbraith *et al.* (1990), one using tints and the other using tincles.

8.4.1 Two-component mixtures of tints

In the isotropic line segment model, suppose that the length probability density function $f(l)$ is of the form

$$f(l) = (1 - p)f_0(l) + pf_1(l).$$

This represents a two-component mixture of long and short tracks, where the density of long tracks f_0 has mean μ_0, the density of short tracks f_1 has mean $\mu_1 < \mu_0$, and p is the proportion of uranium atoms (by volume) giving rise to tracks in the short component. Then the length-biased density $g(l)$ will also have two components, but with a smaller proportion q in the short component, where

$$q = \frac{p\mu_1}{p\mu_1 + (1 - p)\mu_0}. \tag{8.3}$$

Galbraith *et al.* (1990) tested this theory using the Loaded Dog experiments. In Figure 8.5 it can be seen which tracks belong to which component with very few ambiguities, so one may calculate the observed proportions of tracks from the short component. Table 8.2 shows these proportions along with the nominal values of μ_0, μ_1, p and of q given by (8.3). The observed proportions of short tracks are clearly less p and in good agreement with q, as would be expected for length-biased sampling.

This argument is slightly over-simplified because equation (8.3) is based on the isotropic model which is incorrect for the annealed samples. Under the anisotropic line segment model the length-biased density of l given ϕ is given

Table 8.2 *Summary data from the Loaded Dog two-component mixtures.*

Mixture			Number of tints			Proportion		
μ_0	μ_1	p	long	short	total	short	(s.e.)	q
16	10	0.938	14	154	168	0.916	(.021)	0.904
16	8	0.938	15	62	77	0.805	(.045)	0.882
16	10	0.833	39	136	175	0.777	(.031)	0.758
16	8	0.833	24	77	101	0.762	(.042)	0.714

by equation (8.2) so the above theory would apply to tracks at the same angle. However, if one repeats the argument but restricting attention to narrower ranges of angles over which the mean length does not vary significantly, it is still found that the proportions of short tracks agree well with q and not with p. So we have a fairly convincing demonstration that length-biased sampling does apply to tints in practice.

8.4.2 Angles of tincles to host fractures

Length-biased sampling of tincles was also tested by Galbraith *et al.* (1990), arguing as follows. If tincles were sampled randomly from all horizontal tracks, then under the isotropic line segment model their angles ψ to the fracture would have a uniform distribution. But if they were sampled with probability proportional to length, then the joint density function of l and ψ for horizontal tincles becomes

$$l \sin \psi f(l)/\mu, \quad 0 < \psi < \pi/2, \ 0 < l < \infty.$$

The argument here is the same as that leading to (8.1), except that now $l \sin \psi$ is the length-bias factor for tincles at angle ψ. For example, in Figure 8.3 all tracks have the same angle ψ to the fracture, but if they were at different angles, different numbers would intersect and be revealed. These numbers would be proportional to $l \sin \psi$, the length of the track in the direction perpendicular to the fracture.

It follows from the above equation that, for tracks that intersect the fracture, the angle ψ is independent of l and has probability density equal to $\sin \psi$, for $0 < \psi < \pi/2$, regardless of the form of $f(l)$. So one can test the length-bias principle by measuring angles ψ and comparing their observed distribution with this theory.

Figure 8.6 shows a frequency histogram of measurements of ψ for 122 tincles in a sample of Durango apatite. The sample was only lightly annealed and the length distribution was approximately isotropic. The observed distribution of ψ departs markedly from uniform and broadly follows a sine curve (shown by the solid line), though it is even more biased towards high angles.

This suggests that length-biased sampling of tincles can explain the shape of the observed distribution of ψ to a large extent, but not completely. A likely additional factor is *fracture-thickness bias* which we discuss in Section 8.5.

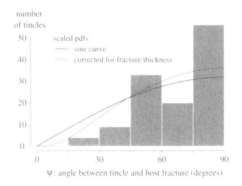

Figure 8.6 *Histogram of the angles ψ of tincles to their host fractures for 122 tincles in a lightly annealed sample of Durango apatite (from Galbraith et al., 1990).*

After adjusting for this, the length-biased sampling distribution of ψ is given by the dotted curve in Figure 8.6 which is in better agreement with the data.

8.5 Fracture-thickness bias

An observer needs to see both ends of a track to be sure that it is horizontal and fully confined, so he or she will only measure tincles that have both ends showing. Thus tincles are subject to a *fracture-thickness bias*, in which longer tracks intersecting a fracture have a higher probability of having both ends exposed, and hence of being measured. A similar bias favours tracks at a high angle to the fracture. In extreme cases, very short tracks may be completely hidden in wider fractures.

We may use the line segment model to calculate approximately the effect of fracture thickness on the observed distribution of ψ, the angle between the horizontal tincle and its host fracture. Figure 8.7 shows an idealised fracture of width β and a horizontal cross-section of a parallelepiped of height $l \sin \psi + \beta$ intersecting a unit area of the fracture. If the centre of a horizontal track of length l at angle ψ to the fracture falls in the parallelepiped, a tincle will be created. But it will not be measured if its centre falls in either shaded region, for then one of its ends falls in the fracture. Hence the expected number of tincles of length l and angle ψ per unit area of the host fracture is proportional to the volume of the parallelepiped minus the shaded ends, which is $l \sin \psi - \beta$.

Therefore, if a host fracture has thickness β, the expected number of such tracks is proportional to $l \sin \psi - \beta$, provided this quantity is positive. Thus the conditional probability density function of ψ given l and β for sampled tincles is

$$\frac{l \sin \psi - \beta}{l\sqrt{1 - (\beta/l)^2} - \beta\{\frac{\pi}{2} - \arcsin(\beta/l)\}}, \quad \arcsin(\beta/l) < \psi < \frac{\pi}{2}.$$

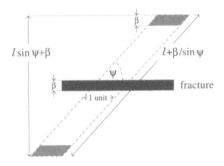

Figure 8.7 *A horizontal cross-section of a parallelepiped of height $\beta + l\sin\psi$ intersecting a unit area of a fracture of width β (from Galbraith et al., 1990).*

For the tincles in Figure 8.6 the fracture widths were measured along with the track lengths and angles ψ. Then a separate conditional density function was calculated for each tincle by using the measured values of l and β. Summing these density functions for all tincles and rescaling gave the dotted curve "corrected for fracture thickness" in Figure 8.6, which, as we noted earlier, is in better agreement with the data.

This analysis ignores other features encountered in measuring tincles. Some fractures cast a shadow which may obscure a track end, particularly in transmitted light. If the width of shadow is s, then the expected number of horizontal tincles with both ends showing is proportional to $l\sin\psi + \beta - 2s$, which reduces to the previous formula when $s = \beta$. Secondly, an observer needs to see enough of a tincle to measure it: if it is at an acute angle to the fracture, so that each end is just visible in theory, it still may not be measurable — indeed, it may be etched so as to merge with the fracture. On present evidence therefore, the effects of fracture thickness and measurement limitations can explain the observed departure from length-biased sampling of tincles.

The above argument assumes implicitly that fractures are formed after the fission tracks are created, and their formation does not alter the length of the intersecting tracks. Fractures, like tracks, are widened by etching, so that this assumption is at least partly correct. Other models of fracture formation would lead to different corrections for fracture thickness.

8.6 Orientation bias

In Figure 8.2 the observed distribution of angles of tints to the c-axis is not uniform. Table 8.3, from Galbraith *et al.* (1990), shows some more extensive data. These are numbers of horizontal tints classified by angle ϕ for five samples of induced tracks in Durango apatite. One sample was unannealed and the others were heated for 1 hour at the temperatures shown.

Table 8.3 *Numbers of horizontal tints at various angles to the c-axis for five samples of Durango apatite (data from Green et al., 1986, 50 s etching).*

Sample	Angle ϕ (degrees) between track and c-axis						Total
	0–15	15–30	30–45	45–60	60–75	75–90	
unannealed	5	18	31	42	55	49	200
260°C	10	23	26	50	47	44	200
310°C	7	22	36	54	44	39	202
336°C	8	24	40	55	41	32	200
352°C	2	17	24	31	27	17	118

Theory of the line segment model suggests that, for horizontal confined tracks, ϕ should be uniformly distributed, and for observed horizontal confined tints, ϕ should be biased towards low angles because of length-biased sampling. But the data clearly depart from this. The frequency distributions have relatively few tracks at low angles and rise to a mode somewhere between 45° and 90°. Also, the mode is at a lower angle for the more heavily annealed samples. For the unannealed sample the range 60–75° has the most tracks whereas for the 310°C and 336°C samples the largest frequencies are for the 45–60° range. Similar features can also be seen in Figure 8.2.

One explanation may lie in the shape of host semi-tracks. When a semi-track is etched, the etchant travels further parallel to the c-axis than perpendicular to it, producing a thin knife blade shape for the etched semi-track (Gleadow, 1981). Thus a potential host semi-track presents a wider face to the direction perpendicular to the c-axis, and this favours the sampling of tints at high angles ϕ.

We can use the line segment model to quantify this effect. If $N(\phi)$ denotes the number of horizontal tracks at angle ϕ, then, to a first approximation,

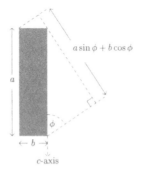

Figure 8.8 *Idealised horizontal cross-section of an etched semi-track. The dimension a (parallel to the c-axis) is about 3.5 to 4 times longer than b. To a potential tint at angle ϕ it presents a face of width $a \sin \phi + b \cos \phi$ (from Galbraith et al., 1990).*

Table 8.4 *Further analysis of the data in Table 8.3*

Angle ϕ between track and c-axis (degrees)

	0–15	15–30	30–45	45–60	60–75	75–90
			fitted frequencies			
unannealed	16.63	26.33	34.12	39.43	41.95	41.54
260°C	16.95	26.76	34.47	39.51	41.60	40.71
310°C	17.55	27.60	35.29	40.02	41.55	39.99
336°C	18.60	28.95	36.29	39.93	39.81	36.42
352°C	15.54	23.16	26.45	24.70	18.52	9.64
			standardised residuals			
unannealed	−2.48	−1.31	−0.42	0.32	1.53	0.89
260°C	−1.35	−0.57	−1.17	1.26	0.65	0.40
310°C	−2.12	−0.84	0.09	1.66	0.30	−0.12
336°C	−2.06	−0.73	0.47	1.78	0.15	−0.59
352°C	−3.29	−1.06	−0.39	0.97	1.42	1.58

the expected number $E[N(\phi)]$ will be proportional to their mean length $\mu(\phi)$ multiplied by the width of the host track in the direction perpendicular to ϕ. From Figure 8.8 this width is $a\sin\phi + b\cos\phi$, where a and b are the horizontal cross-sectional dimensions of the host semi-track. Thus

$$E[N(\phi)] \propto (a\sin\phi + b\cos\phi)\mu(\phi).$$

As ϕ varies from 0 to 90°, this curve typically rises to a maximum, at a point depending on the ratio a/b and on how fast $\mu(\phi)$ varies, and then drops. Thus for a lightly annealed sample, $\mu(\phi)$ is nearly constant and $E[N(\phi)]$ will peak at a high angle, while for a more heavily annealed sample, $\mu(\phi)$ drops more rapidly so that $E[N(\phi)]$ will peak at a lower angle. So this model behaves qualitatively in the right way to agree with the data.

Table 8.4, from Galbraith *et al.* (1990), gives some fitted frequencies and standardised residuals based on this model. These used the parametric form $\mu(\phi) = \eta + \xi\cos\phi$ and $a/b = 3.7$ from independent measurements of the dimensions of etched openings of semi-tracks on a prismatic face.

The model fails to explain the data completely, especially at low angles where too few tracks are seen, but it does explain much of the broad pattern. Excluding angles below 30° and re-fitting the model produced a good fit. This paucity of tints at low angles to the c-axis is commonly encountered by observers. Whether the correct explanation lies in a further sampling bias, an etching effect or some other cause is perhaps an open question. Donelick *et al.* (1999) presented data showing similar non-uniform distributions of angles between horizontal tints and the c-axis.

For unannealed samples, the preferential sampling of tints at high angles should not affect the observed length distribution, because l does not depend on ϕ. But for annealed samples, it could do. Thus the length distribution of annealed tints is affected by orientation bias. This favours tracks at a high angle ϕ and these tracks are relatively short, so orientation bias acts in the opposite direction to length bias. For tincles, the angular distribution will depart from uniform if host fractures tend to run in a particular direction and this could be reflected in the measured lengths.

A simple way to correct for orientation bias is to measure the angle ϕ between each horizontal track and the c-axis, and standardise all distributions to have a common distribution for ϕ, for example, the uniform distribution between 0 and $\frac{\pi}{2}$.

Also, it is wise to separate tints from tincles when measuring confined track lengths. Calculation of mean tint length and mean tincle length in routine work has shown that the latter is slightly greater in general, particularly in annealed samples (e.g., Table 8.1). This is a consequence, at least partly, of observed differences in orientation bias between tints and tincles, but it could also be attributed to fracture-thickness bias in sampling tincles, which we discussed earlier. Anisotropic etching implies that fractures at a high angle to the c-axis will tend to be wider, and this is also seen in practice.

8.7 Surface proximity bias

A short track that is nearly horizontal and lying close to the polished surface or to the edge of the grain is more likely to be fully confined than a longer one. Figure 8.9 illustrates this. Tracks a and b are produced from atoms the same short distance from the surface and have the same small dip angle. Track a may be identified as a horizontal confined track, but the longer track b will not as it has intersected the surface and become a semi-track. Tracks c and d have the same lengths and dip angle as a and b, but are further from the surface and are both fully confined.

Thus a sample of horizontal confined tracks emanating from atoms close to the surface will be biased towards shorter tracks. Of course in order to be etched, a confined track must also intersect a fracture or an etched track, which it is more likely to do if it is longer. So surface proximity bias and length bias act in opposite directions.

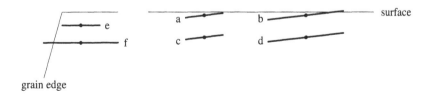

Figure 8.9 *Surface proximity bias. A shorter track near to the surface or grain edge is more likely than a longer track to be totally confined.*

Table 8.5 *Maximum likelihood estimates of the mean track length μ (in microns) from tincles and tints in the Loaded Dog control samples. The numbers of tracks n and mean lengths $\bar{y}$ are from Table 8.1.*

		n	$\bar{y}$	$\hat{\mu}$	se($\hat{\mu}$)
Unannealed	Tincles	27	16.14	15.92	0.17
	Tints	173	15.79	15.75	0.06
Annealed, 300°C	Tincles	57	10.13	9.88	0.12
	Tints	203	9.97	9.75	0.04
Annealed, 343°C	Tincles	58	8.27	6.90	0.14
	Tints	166	7.36	6.59	0.07

A similar bias favours short tracks near the grain edge. In Figure 8.9, track e is fully confined, but the longer track f, which is produced by an atom the same distance from the edge, is not.

8.8 Estimates of μ from horizontal confined tracks

To see some numerical implications of these observational effects, Table 8.5 shows estimates of the underlying mean length μ from the measurements of tints and tincles in the Loaded Dog control samples. The estimates have been corrected for orientation bias by standardising to the same distribution of ϕ. They allow for length-biased sampling, and for tincles they also allow for fracture thickness bias. Estimation is by maximum likelihood using the parametric model for E[$l|\phi$] in equation (7.53). Details of the likelihood function are discussed in Chapter 9. Table 8.5 also gives the numbers of tracks measured n and the unweighted mean lengths $\bar{y}$ from Table 8.1.

The estimates $\hat{\mu}$ are less than $\bar{y}$ in all cases, the differences being greater for the more heavily annealed samples. It is still true that the estimates from tincles are slightly longer than those from tints, but they are in closer agreement after adjusting for orientation and fracture-thickness bias. Length-biased sampling makes little difference to the mean length estimates for single populations such as these, but it can have a substantial effect for mixed populations.

8.9 Projected semi-track lengths and angles

Information about the underlying the length distribution $f(l)$ can also be obtained by measuring semi-tracks — those parts of tracks that have intersected the plane of observation and so have been etched.

The simplest measurements are to record the coordinates of the semi-track ends on the plane of observation, and the direction of the c-axis, in the same way as for confined tracks. This allows one to calculate the *projected* lengths and angles, r and ω with respect to the coordinate system in Figure 7.1. It is possible to measure also the depth of the semi-track end below the plane,

Table 8.6 *Number of semi-tracks* n, *mean* $\bar{r}$ *and standard deviation* s_r *in microns, and coefficient of variation* $s_r/\bar{r}$ *of projected semi-track lengths in three samples of Durango apatite from the Loaded Dog experiments.*

	n	$\bar{r}$	s_r	$s_r/\bar{r}$
Unannealed	496	5.16	3.55	0.69
Annealed, 300°C	491	3.44	2.40	0.70
Annealed, 343°C	495	2.67	2.30	0.86

and hence to calculate the actual semi-track length t and angle ϕ, but this is usually considered impractical for routine analysis.

Semi-tracks are more numerous in a field sample than are confined tracks, but they also give much less direct information about $f(l)$, for several reasons. Firstly, they intersect the plane of observation from a variety of depths and at a variety of angles, so that, even if all latent tracks had the same length l, their projected lengths r will still vary between 0 and l. Secondly, the full track is not revealed because part of it has been polished away. And thirdly r is not the full length of the remaining part, but merely its projection onto the plane. In addition, because a long track is more likely to intersect a surface than a short track, the observed sample will be biased towards long tracks. In Chapter 7, we derived formulae for the theoretical probability distributions of r and t in terms of $f(l)$, the probability distribution of latent track lengths l, which take into account these features. Some example probability density functions are plotted in Figure 7.5. Formulae relating the means and variances of these distributions are also given in Chapter 7.

Table 8.6 gives the sample means and standard deviations of r for about 500 semi-tracks measured in each of the Loaded Dog control samples. Data from confined tracks in these samples are shown in Figure 8.2 and in Table 8.1. As expected from theory, the mean projected lengths are roughly one third of the mean confined lengths for the same samples, and the coefficients of variation are about 70% or higher and increase with the amount of annealing. Recall from Section 7.1 that the coefficient of variation of a triangular distribution is 71%.

Figure 8.10 shows scatter plots and histograms r and ω for two of these samples along with one of the mixture samples, where each grain has a mixture of short and long tracks in the ratio of about 5:1. For the unannealed sample, where the confined track lengths have a small standard deviation, the projected length distribution is nearly triangular, and there is no apparent relation between r and ω. For the annealed sample, though, there are relatively more short projected lengths and few longer projected lengths at high angles. The third sample is a 5:1 mixture of unannealed and annealed tracks. The scatter plot and histogram look like mixtures of those from the two other samples; there is some anisotropy but also a number of longer projected lengths at high angles.

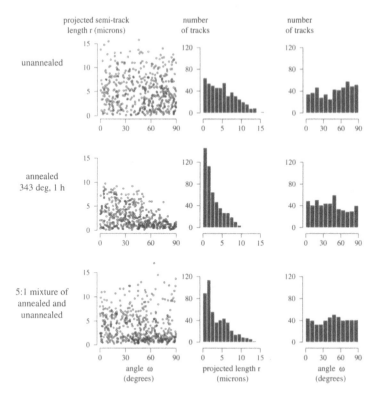

Figure 8.10 *Scatter plots and histograms of r and ω for three samples of induced fission tracks in Durango apatite from the Loaded Dog experiments: an unannealed sample, an annealed sample and a 5:1 mixture of annealed and unannealed tracks.*

Measurements of r have several general features. Firstly, their distribution has a high dispersion even for clean samples, such as a set of induced tracks, or spontaneous tracks from a field sample in which negligible annealing has taken place. Secondly, their frequency distribution is determined largely by geometrical properties — the operations of truncation and projection — and is insensitive to the form of $f(l)$, so that simple inspection of the frequency histogram will not easily reveal important characteristics of the thermal history. And thirdly, samples of projected length measurements often have a lack of short lengths compared with theoretical expectation, which complicates their interpretation.

Although they are less informative than confined track lengths, projected lengths are nevertheless potentially useful. Firstly, horizontal confined tracks may be scarce, because of low uranium content, a scarcity of host fractures or heavy annealing. Secondly, the biases in projected lengths are essentially geometrical and are easier to take into account because there is less subjective selection when sampling semi-tracks. Thirdly, a large number of projected lengths can be measured quickly, so lack of information per track might be compensated for by measuring larger numbers. And fourthly, even when confined tracks are present, individual grains may have none, so that projected lengths may provide information on a grain-by-grain basis. It is therefore useful to understand what can be inferred from them.

The potential information in measurements of r and ω was studied in detail by Laslett et $al.$ (1994) in the context of estimating the mean track lengths μ_0 and μ_1 and the mixing proportion p in a two-component mixture (see Section 7.9). For a thermal history like that in panel (d) of Figure 1.3, these parameters relate to the current temperature, the maximum palaeo-temperature and the timing of uplift and erosion or cooling. It was found that it is not feasible to estimate all three parameters reliably from samples of up to 2000 semi-tracks. However, a good estimate of μ_0 can usually be obtained from knowledge of the geology and current thermal environment. Then measurements of r and ω can provide useful estimates of μ_1 and p.

More specifically, to obtain reasonably precise estimates, p needs to be fairly large (at least 0.6 and preferably 0.8 or higher) corresponding to relatively recent cooling; and μ_1 needs to be quite small, corresponding to cooling from a high temperature. In these favourable cases, sample sizes of over 500 semi-tracks are needed to get precise estimates, although useful constraints on the thermal history may be obtainable with lesser numbers. By contrast, estimates from rather fewer confined track measurements are much more informative over a wide range of μ_1 and p. Furthermore, measurements of the angle ω, in addition to r, do contain extra information. For a heavily annealed sample from a single component, or a two-component mixture with a high proportion of short tracks, the information obtained from ω increases the precision of the estimate of μ_1 but adds little to the estimate of p.

Projected semi-track measurements can also provide useful estimates of a single parameter, such as a single mean length, or a mixing proportion when both component mean lengths are known. Table 8.7 shows the results of using the measured projected semi-track lengths and angles (r, ω) to estimate the underlying mean length μ, for each of the Loaded Dog control samples. These are maximum likelihood estimates taken from Laslett et $al.$ (1994), which we may compare with the estimates from horizontal confined tracks in Table 8.5.

Table 8.7 *Maximum likelihood estimates of the mean track length μ in microns from measurements of projected semi-track lengths r and angles ω for the Loaded Dog control samples (from Laslett et al., 1994). Numbers of semi-tracks n, and their mean projected length $\bar{r}$ are from Table 8.6.*

	n	$\bar{r}$	$\hat{\mu}$	se($\hat{\mu}$)
Unannealed	496	5.16	15.12	0.37
Annealed, 300°C	491	3.44	9.51	0.26
Annealed, 343°C	495	2.67	6.55	0.20

The estimates from projected semi-tracks are slightly shorter than those from confined tracks, but are formally consistent with them. A rough method of estimating μ is to multiply $\bar{r}$ by 3. This is the "method of moments" estimate based on the single length model in Section 7.1. It can be seen that this gives poor estimates, especially for the annealed samples.

8.10 Semi-track lengths and angles

Laslett and Galbraith (1996a) used a simulation study to assess the potential information in the actual semi-track lengths t and angles ϕ for estimating the parameters p and μ_1. The conclusions paralleled those for projected semi-track measurements, though measurements of t were understandably more informative than r for estimating p and μ_1. In general, samples sizes of at least 100 tracks were necessary to obtain useful estimates and at least 500 tracks were needed to achieve relative standard errors below 10%. Again, measurements of the orientation (in this case ϕ) improved the estimate of μ_1 somewhat but made only a small improvement to the estimate of p.

This study also quantified the improvement in precision gained from using actual semi-track lengths and angles, rather than their projections onto the plane. In those cases where good estimation was possible, it needed about four times as many measurements of r and ω to achieve the same precision as that obtained from t and ϕ.

Another statistical message to emerge from both of these studies was that simple estimates based on sample moments are considerably less efficient than maximum likelihood estimates. For example, when all tracks have the same length l, the probability distribution of t is uniform between 0 and l, with mean $\frac{1}{2}l = \frac{1}{2}\mu$. So a simple estimate of μ would be $2\bar{t}$, where $\bar{t}$ is the sample mean of t. But it is well known that there are much more efficient estimates than this for the parameter of a uniform distribution. The likelihood function is determined by the sufficient statistic, which is the maximum observed t, not the sample mean. In reality, the distribution of t is not exactly uniform, but it is nearly uniform over much of its range and it is still true that moment-based estimates are quite seriously sub-optimal. The same also applies to measurements of projected lengths r. This argues for the use of likelihood methods when estimating parameters of the distributions of t and r.

8.11 Bibliographic notes

This chapter is largely based on material from Galbraith *et al.* (1990), Laslett *et al.* (1994) and Laslett and Galbraith (1996a). Cox (1969) gave formulae for estimating distributions from length-biased samples. Barbarand *et al.* (2003a) reported a detailed empirical study of fission track length and angle measurements in apatite. This compared different observers, apatites of different chemical composition, and different amounts of annealing.

CHAPTER 9

Further developments

This chapter brings together ideas from earlier chapters and points to some statistical aspects of annealing models and thermal history estimation. Much of the methodology outlined here is the subject of ongoing research, and there is considerable scope for further development of the statistical modelling and inference. Sections 9.1 and 9.2 consider the estimation of a two-phase thermal history, using the combined likelihood of counts, lengths and orientations of semi-tracks, and of lengths and orientations of confined tracks. These draw on several earlier sections, notably 2.5, 3.1, 5.1, 7.9, 8.2, 8.5, 8.6 and 8.9. Sections 9.3 to 9.6 briefly discuss the design and analysis of laboratory annealing experiments, while Sections 9.7 to 9.9 outline the steps involved in estimating thermal histories from field data.

The likelihood function is basic to the statistical modelling of fission track data and the estimation of thermal histories. Given this function, there are standard recipes for estimating parameters and their precisions, and for testing which models are well supported by the data. Such estimates and tests have optimal statistical properties under general conditions. It is also straightforward to combine information from different types of data — something that other criteria, such as weighted least squares, are not suited to. Furthermore, the theory of the line segment model enables us to specify the relevant parameters and probability distributions with some confidence.

9.1 Thermal history parameters

Routine fission track dating, as presented in Chapter 3, uses equation (3.6) or (3.10) to estimate the age t of a crystal, usually interpreted as the time since cooling shortly after its formation. But, for illustration, consider the more complex history illustrated in Figure 9.1. Suppose that newly formed apatite crystals are deposited in a sedimentary basin at a time t_0 million years ago. They are progressively buried and, with increasing depth below the earth's surface, there is a corresponding rise in temperature. Then at time t_u, when the temperature is T_u, the basin is uplifted and there is rapid erosion. After this, there is further slow burial until the present day, when the temperature is T_c, say. In practice the current temperature T_c will be known, or can be measured directly, but the other quantities may be unknown. We want to estimate, simultaneously with t_0, the time t_u and maximum temperature T_u at uplift. Because spontaneous tracks have been heated, we are going to need the full fission track age equation (3.5) rather than (3.6).

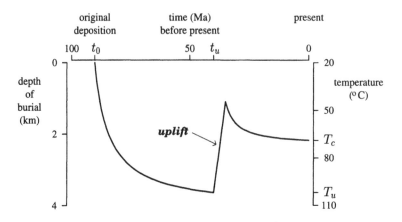

Figure 9.1 *A thermal history producing a two-component mixture of track lengths.*

To relate these parameters to fission track measurements, we need to know how track lengths depend on the temperature-time paths that they have experienced. Although this subject is still developing, the general nature of this relationship is well understood and is illustrated in Figure 9.2, which is based on laboratory annealing experiments using apatite, extrapolated to geological time scales at lower temperatures. The left panel shows how tracks held at a fixed temperature shorten over time. Their rate of shortening decreases with time and becomes very slow. The right panel shows how tracks heated for a fixed time shorten with increasing temperature. Here, as temperature increases, the rate of shortening increases and becomes very rapid. In particular, the rise in temperature needed to completely anneal tracks that have been reduced to about 6 or 7 microns in length is only a few degrees C. These two things together imply that the final length of a track is largely determined by the *maximum* temperature it has experienced throughout its history.

Spontaneous fission tracks form throughout a crystal's history. Those formed between t_0 and t_u will anneal as the temperature rises. The lengths of those formed later within this period will shorten more rapidly and, after a few million years will be practically indistinguishable in length from those formed earlier. Tracks formed after t_u are annealed only slightly and these too will all have similar lengths at the present day ($t = 0$). Hence, a crystal experiencing the thermal history in Figure 9.1 will contain, to a good approximation, a two-component mixture of tracks: a proportion p formed prior to uplift, having mean length μ_1 and a proportion $1 - p$ formed subsequently, having mean length μ_0, where

$$1 - p = \frac{p(t_u)}{p(t_0)} = \frac{1 - e^{-\lambda t_u}}{1 - e^{-\lambda t_0}} \approx \frac{t_u}{t_0}.$$

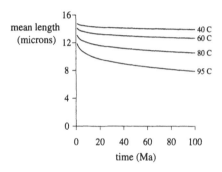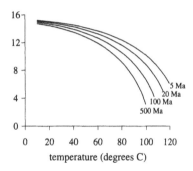

Figure 9.2 *The nature of the dependence of mean fission track length on time and temperature, based on laboratory annealing models for apatite extrapolated to geological time scales at lower temperatures. Temperatures and times are indicative only.*

This type of mixture differs from that considered in Chapter 5. There each grain came from one of two populations. Here each track comes from one of two populations and each grain may contain tracks from both.

Assume for the moment that the relationship between mean fission track length and temperature is known. We return to this aspect briefly in Sections 9.5 and 9.6. Then μ_0 may be inferred from the current temperature T_c. Furthermore, if we estimate μ_1 we can infer T_u. Also from the above equation, if we estimate t_0 and p we can then infer t_u.

9.2 Combined likelihood for track measurements

Let us formulate the likelihood function for the three parameters t_0, p and μ_1. It is sufficient to derive this for data from a single crystal grain. The log likelihood from several grains that have had the same history is obtained simply by adding up that for each grain. Using the notation in Chapters 3 and 8, the data for a single grain may consist of

(a) the number N_s of spontaneous tracks intersecting an polished area A of crystal, and the number N_i of induced tracks intersecting the corresponding area of the mica (using the external detector method);

(b) measured lengths and angles to the c-axis (l_j, ϕ_j), $j = 1, 2, \ldots, n_t$, for n_t horizontal confined tints;

(c) measured lengths, angles to the c-axis, fracture widths and angles to the fracture $(l_j, \phi_j, \beta_j, \psi_j)$, $j = 1, 2, \ldots, n_c$, for n_c horizontal confined tincles; and

(d) measured projected semi-track lengths and angles (r_j, ω_j), $j = 1, 2, \ldots, n_s$, for n_s semi-tracks. Usually these will be the same tracks that are counted in (a), with $n_s = N_s$.

Not all of these may be present, or need be present, for each grain. In fact, each type of data contributes independently to the joint log likelihood, which is the sum of the separate log likelihoods for each data type.

9.2.1 The likelihood for counts

From Chapters 2 and 3, the track counts N_s and N_i have independent Poisson distributions with means $A\rho_s$ and $A\rho_i$, where

$$\rho_s = \tfrac{1}{2}\left(\frac{e^{\lambda t_0}-1}{\lambda}\right)\lambda_f T_c\,\mu_s \quad \text{and} \quad \rho_i = \tfrac{1}{4}\Phi\sigma_f I T_c\mu_i$$

as in equations (3.2) and (3.4). Here μ_s is the *equivalent isotropic length* of the spontaneous tracks, which will differ from μ_i because some of these tracks have been heated. We write $\mu_s = \mu^*$ to emphasise this, where μ^* is given by (2.17). A good estimate of μ_i, the mean length for unannealed induced tracks, should be available from laboratory studies. Taking the ratio ρ_s/ρ_i and substituting $\zeta\rho_d = \Phi\sigma_f I/\lambda_f$ gives the full fission track age equation for t_0, viz.,

$$t_0 = \frac{1}{\lambda}\log\left(1+\frac{1}{2}\lambda\zeta\rho_d\,\frac{\rho_s}{\rho_i}\,\frac{\mu_i}{\mu^*}\right),\tag{9.1}$$

which is the same as (6.1).

As in Chapters 5 and 6, to eliminate the nuisance parameter T_c we may use the conditional probability distribution of N_s given N_s+N_i, which is binomial with index N_s+N_i. From Section 5.1, equation (5.10), the log likelihood from the track counts is therefore

$$L_a = N_s\log\theta + N_i\log(1-\theta) + \text{constant},\tag{9.2}$$

where

$$\frac{\theta}{1-\theta} = \frac{\rho_s}{\rho_i} = \frac{2}{\zeta\rho_d}\left(\frac{e^{\lambda t_0}-1}{\lambda}\right)\frac{\mu^*}{\mu_i}.$$

Also, from Section 7.9,

$$\mu^* = (1-p)\mu_0^* + p\mu_1^*,$$

where μ_0^* and μ_1^* are the equivalent isotropic lengths of tracks from the two separate components. Therefore θ, and hence L_a, is a function of the three unknown parameters t_0, p and μ_1^*, which are not separately identifiable without further information.

9.2.2 The likelihood for tints

Suppose that there are n_t tints intersecting semi-tracks and that their lengths and angles (l,ϕ) have been measured. Numbers and orientations of tints contain thermal history information, but Section 8.6 suggests that this is confounded with unknown sampling biases, particularly orientation bias. It is therefore sensible to use the likelihood from the conditional distribution of length given orientation. Allowing for length-biased sampling, the probability density of this conditional distribution is given by (8.2), *viz.*,

$$g_\phi(l) = l f_\phi(l)/\mu(\phi), \quad 0 < l < \infty, \tag{9.3}$$

where $\mu(\phi) = \mathrm{E}[l|\phi]$. In the present case, $f_\phi(l)$ may be written as

$$f_\phi(l) = (1-p) f_{\phi 0}(l) + p f_{\phi 1}(l) \tag{9.4}$$

as in equation (7.46) and

$$\mu(\phi) = (1-p)\mu_0(\phi) + p\mu_1(\phi) \tag{9.5}$$

in the obvious notation.

In a fully parametric approach, one might assume that each component distribution $f_{\phi 0}$ and $f_{\phi 1}$ has a specific form — for example, $f_{\phi 1}(l)$ may be a Gaussian probability density with mean and $\mu_1(\phi)$ and variance $v_1(\phi)$ defined in terms of a small number of parameters, and similarly for $f_{\phi 0}(l)$. A simple parametric form for $\mu_1(\phi)$ is

$$\mu_1(\phi) = \eta_1 + \xi_1 \cos\phi,$$

where η_1 and ξ_1 may further be defined in terms of the mean length μ_1 by a relation such as (7.53) so that $\mu_1(\phi)$ depends on just one parameter, μ_1. The simplest assumption for $v_1(\phi)$ is that it is constant for all ϕ, equal to σ_1^2, say. If this is unrealistic, a useful parametric form is

$$v_1(\phi) = \sigma_i^2 + \sigma_1^2 \left(1 - \mu_1(\phi)/\mu_i\right)^\kappa,$$

with $\kappa = 1$ or 2. Here μ_i and σ_i are the mean and standard deviation of unannealed induced track lengths. Good estimates of these should be available from laboratory studies. If σ_1 is greater than zero, the second term increases both with angle and with the amount of annealing. Other approaches are possible, but with a parametric model it is much easier to account for sampling and observational biases.

Hence the log likelihood from sampled tints is

$$L_b = \sum_{j=1}^{n_t} \log\left(l_j f_{\phi_j}(l_j)/\mu(\phi_j)\right) + \text{constant}, \tag{9.6}$$

with $\mu(\phi)$ given by (9.5). From (9.4), $f_\phi(l)$ and hence L_b is a function of p and μ_1, and of any variance components, such as σ_0 or σ_1, that are not otherwise estimated independently.

9.2.3 The likelihood for tincles

Similar considerations apply to tincles. The probability of sampling a tincle of a particular length and orientation depends on the number, orientations and sizes of the fractures in the crystal. It is therefore important to measure these for the actual host fractures and condition on them to avoid terms involving unknown or unmeasurable quantities. From Section 8.5 the conditional probability density of l given ϕ, β and ψ, allowing for length bias and fracture-thickness bias is

$$g(l|\phi, \beta, \psi) = \frac{(l_j - \beta_j / \sin \psi_j) f_{\phi_j}(l_j)}{\int_{\beta_j / \sin \psi_j}^{\infty} (l - \beta_j / \sin \psi_j) f_{\phi_j}(l) \, dl}, \tag{9.7}$$

where the denominator is the normalising quantity required to ensure that probability density function integrates to 1. Thus if (l, ϕ, β, ψ) are measured for each of n_c tincles, the conditional likelihood of the lengths given the other measurements is

$$L_c = \sum_{j=1}^{n_c} \log \left(g(l_j | \phi_j, \beta_j, \psi_j) \right) + \text{constant}, \tag{9.8}$$

where g is given by (9.7) with $f_\phi(l)$ as in (9.4). As for L_b, this is a function of p and μ_1 and of any unknown variance components.

The specific forms in (9.7) and (9.3) are not necessarily definitive. These condition on orientation and account for length bias and fracture-thickness bias. If other biases are non-negligible, then a more general bias factor $b_\phi(l)$ might be needed and the observed length distribution would be of the form

$$g_\phi(l) = \frac{b_\phi(l) f_\phi(l)}{\int_0^{\infty} b_\phi(l) f_\phi(l) \, dl}.$$

9.2.4 The likelihood for projected semi-tracks

For projected semi-tracks, both lengths and angles are informative. The log likelihood from these is

$$L_d = \sum_{j=1}^{n_s} \log \left(f_{ps}(r_j, \omega_j) \right) + \text{constant}, \tag{9.9}$$

where from equation (7.47)

$$f_{ps}(r, \omega) = (1 - q) f_{ps0}(r, \omega) + q f_{ps1}(r, \omega),$$

with $q = p\mu_1^*/\mu^*$ and where f_{ps0} and f_{ps1} are given in terms of the underlying conditional distributions of l given ϕ by (7.27). Again, the same parametric forms may be used and L_d is then a function of the same parameters as L_b and L_c.

It is not usual current practice to include measurements from projected semi-tracks, but this can be done in principle and could be useful for the reasons mentioned in Section 8.9.

9.2.5 The combined likelihood

The overall log likelihood is the sum of all contributions, *viz.*,

$$L = L_a + L_b + L_c + L_d$$

for each grain and summed over grains. The parameters t_0, p and μ_1 can be estimated by maximising L.

However, t_0 appears only in L_a in (9.2), as part of the composite parameter θ. From Chapter 5, the maximum likelihood estimate of θ is $N_s/(N_s + N_i)$, where N_s and N_i are the total counts over all grains. So p and μ_1 are estimated from the last three terms, i.e., by maximising $L_b + L_c + L_d$, which of course needs to be done numerically. Hence estimates of μ_1^* and μ^* may be calculated and then t_0 may be estimated from (9.1) by substituting N_s/N_i in place of ρ_s/ρ_i.

9.3 Annealing experiments

Translating track length parameters into temperature information requires a quantitative relationship between fission track length, temperature and time. The form of this relationship is usually obtained from laboratory annealing experiments. These produce data from samples of annealed fission tracks, to which equations are fitted. Such *annealing models* are then used as ingredients in the calculation of thermal histories from field data. There is a growing literature about annealing and thermal history modelling, which is beyond our present scope. Here we mention briefly some of the statistical aspects.

Annealing experiments usually take the following form. Apatite (or other mineral) samples are first heated to a high temperature to remove all fossil tracks. They are then irradiated to create induced fission tracks with a length distribution similar to that for unannealed tracks. One sample is kept aside as a control and other samples are heated at various chosen temperatures for various chosen times to produce a suite of samples in which tracks have undergone differing amounts of annealing. Fission tracks are etched and measured in the usual way for each sample.

There are essential differences between the fission track distributions in such experiments and those in natural field samples. In a laboratory sample, all of the induced tracks are heated to the same temperature for the same time, whereas in field data, spontaneous fission tracks have formed continuously over geological time, and therefore tracks in the same sample may have experienced different maximum temperatures. Also in the laboratory, the times are much shorter and the temperatures much higher than in the field, to achieve a given amount of annealing. These and other differences need to be taken into account when using annealing models to estimate thermal histories. In fact, models derived from laboratory experiments have been quite successful when extrapolated to the longer geological time scales at lower temperatures. Nevertheless, further developments are needed to improve this aspect.

From a statistical viewpoint there are a number of choices to be made. What data should be used? What response variable or variables? What form of model equation? What form of error structure? And what method of fitting models? Also, when designing annealing studies, what temperatures and times should be chosen?

9.4 Annealing data

In early annealing studies, attempts were made to fit models to changes in track density, and then to relate density to track length. This has since been superseded by a more direct approach using measurements of confined track lengths, taking into account the features mentioned in Chapters 7 and 8.

The first extensive laboratory annealing data using confined track lengths were published by Green *et al.* (1986) using Durango apatite. These used 77 annealing experiments for times ranging from 20 minutes to 500 days. Later studies have used a wider variety of apatites, but annealing times of up to 1000 hours (about 42 days) only. These include Crowley *et al.* (1991), using two different fluorapatites, and Carlson *et al.* (1999), Barbarand *et al.* (2003b) and Ravenhurst *et al.* (2003) all using several different apatites including Durango apatite. In the last three of these studies, angles to the *c*-axis were measured in addition to lengths. Annealing data using confined track lengths in zircon were published by Yamada *et al.* (1995) and Tagami *et al.* (1998).

For illustration, let us look at the form of data in Table 9.1, which are taken from Barbarand *et al.* (2003b). There are 25 annealing experiments and a control. Aliquots of several different apatites were heated at specified temperatures for a given time, between 1 hour and 1000 hours. For each, the lengths and orientations of about 100 horizontal confined tracks (tints only) were measured, except when fewer than 100 confined tracks were seen, in which case all were measured.

In some heavily annealed samples, no confined tracks at all were seen. This is a type of informative censoring that usually arises in annealing experiments. When the temperature is high enough, there are no tracks to measure, but this fact itself implies that, at the given temperature and time, the true mean length must be small. For statistical modelling, this requires an equation expressing the probability that no confined tracks are seen as a function of the parameters in the model.

Table 9.1 shows summary data consisting of the number n_i, mean y_i and standard deviation s_i of confined track lengths for the apatites from Durango and Deep River Valley. The former contains about 0.4% chlorine, while the latter is a fluorapatite having only about 0.01% chlorine. In every annealed sample (i.e., except for the control) the mean track length y_i is shorter for the Deep River Valley apatite, reflecting a greater sensitivity to temperature. For aliquots heated for the same time, the mean length also decreases with increasing temperature, with one exception — the Deep River Valley sample heated for 10 h at 320°C. This sample and the next were measured using a technique called "Cf irradiation" whereby artificial semi-tracks are created in the crystal for the etchant to pass down, in order to increase the probability of observing confined tracks. They are created by irradiating crystals with heavy ions ([252]Cf-derived fission fragments) and the confined tracks revealed by intersecting them are effectively tints, but where the host semi-tracks are all parallel. Barbarand *et al.* (2003a) found that in heavily annealed samples

Table 9.1 *Annealing summary data for apatites from Durango, Mexico, and Deep River Valley, Canada (from Barbarand et al., 2003b).*

time	temp ($^\circ$C)	n_i	Durango y_i (μm)	s_i (μm)	y_{ai} (μm)	s_{ai} (μm)	Deep River Valley n_i	y_i (μm)	s_i (μm)
control	20	100	15.77	0.86	15.75	0.89	100	16.05	0.80
1 hour	270	100	14.33	0.91	14.27	0.88	100	13.99	0.80
	320	100	11.74	0.97	11.71	0.89	100	10.63	1.07
	340	100	10.56	1.09	10.30	0.82	85	8.59	2.38
10 hours	200	100	14.66	0.77	14.66	0.76	100	14.50	0.80
	240	100	14.07	0.85	14.10	0.81	101	13.72	0.71
	275	100	12.64	0.79	12.71	0.77	100	12.45	1.11
	280	100	12.71	1.05	12.67	0.92	100	11.43	1.11
	280	100	12.43	0.84	12.46	0.74			
	300	100	11.18	1.05	11.14	0.81	100	9.91	1.37
	310	100	10.28	1.09	10.21	0.89			
	312	102	10.24	1.37	10.13	1.11	100	6.50	2.79
	320	100	8.70	2.30	8.35	1.90	*111*	*7.06*	*2.33*
	325	100	8.69	2.40	7.94	1.94	*100*	*6.39*	*2.23*
	335	0	0	0	0	0	0	0	0
100 hours	210	100	14.41	0.83	14.32	0.75	100	14.09	0.87
	255	100	13.13	0.80	13.02	0.78	100	11.81	0.99
	275	100	11.02	1.25	10.98	1.04	100	9.51	1.44
	287	102	10.34	1.12	10.08	0.98	55	7.59	2.73
	295	100	8.85	1.93	8.08	1.83	0	0	0
500 hours	257	100	11.49	0.96	11.53	0.78	100	8.70	2.55
	266	100	9.83	1.40	9.58	1.20	100	7.06	2.90
1000 hours	185	100	14.36	0.75	14.37	0.68	100	14.29	0.83
	225	100	12.93	0.95	12.77	0.77	100	12.32	0.90
	250	100	11.02	1.07	11.04	0.86	100	10.08	1.24
	266	100	9.71	1.55	9.31	1.33	100	7.39	2.56

Notation: n_i is the number of confined tracks measured in experiment i; y_i and s_i are the mean and standard deviation of track lengths; while y_{ai} and s_{ai} are the mean and standard deviation of lengths adjusted to a common orientation distribution. Measurements in italics used Cf irradiation.

confined tracks revealed by Cf irradiation were on average a bit longer than natural tints. Some allowance therefore needs to be made for this.

9.5 Annealing models

Consider a cohort of tracks that have been annealed at temperature T for time t. An annealing model equation expresses the mean or expected length μ of such tracks as a function of T and t. The physics and chemistry of annealing are not well enough understood to determine an appropriate mathematical form of such an equation from physical considerations alone. For example it is clear that first order kinetics, where the rate of annealing is proportional to the amount of annealing, is unsatisfactory (Green *et al.*, 1988). Models in current use therefore aim to express the main observed features of annealing data in terms of a few parameters. These have traditionally been based on the Arrhenius plot of $\log t$ against $1/T$, where T is absolute temperature (Kelvin).

A general class of such models is of the form

$$\log t = A(r) + B(r)/T,\qquad (9.10)$$

where $A(r)$ and $B(r)$ are suitable functions of the length *reduction* r, i.e., the factor by which the mean track length has been reduced from that for unannealed tracks. In equation (9.10), a contour of constant r, and hence of constant μ, is a straight line on the Arrhenius plot with slope $B(r)$.

If $B(r)$ does not depend on r, the contour lines are all parallel and a relatively tractable class of models is obtained. An example of such a *parallel* model is

$$\mu = \mu_{\max}\left(1 - \exp\{c_0 + c_1(\log t - b/T)\}\right)^{\lambda},\qquad (9.11)$$

where t is time (usually in hours) and T is temperature in Kelvin. Writing $r = \mu/\mu_{\max}$, this may be expressed as

$$\log(1 - r^{1/\lambda}) = c_0 + c_1(\log t - b/T)\qquad (9.12)$$

and is of the form (9.10) with $B(r) = b$ and $A(r) = [\log(1 - r^{1/\lambda}) - c_0]/c_1$. The quantities $\mu_{\max}$, λ, c_0, c_1 and b are parameters to be determined. However, parallel models have not been generally successful for fitting data and there is now fairly strong empirical evidence that $B(r)$ increases with the amount of annealing, at least within the class of models (9.10).

A model equation that has been used very successfully is

$$\mu = \mu_{\max}\left[1 - \exp\left\{c_0 + c_1\frac{\log t - \log t_c}{1/T - 1/T_c}\right\}\right]^{\lambda},\qquad (9.13)$$

where $\mu_{\max}$, λ, c_0, c_1, T_c and t_c are unknown parameters. In terms of $r = \mu/\mu_{\max}$ this may be expressed as

$$\log(1 - r^{1/\lambda}) = c_0 + c_1\frac{\log t - \log t_c}{1/T - 1/T_c}.\qquad (9.14)$$

This is an example of a *fanning* model. It is of the form (9.10) where the contour lines all meet at the point $(1/T_c, \log t_c)$ and their slopes increase as μ decreases. Both $A(r)$ and $B(r)$ are now linear functions of $\log(1 - r^{1/\lambda})$. For a sensible model, the point $(1/T_c, \log t_c)$ needs to be far below the values of $(1/T, \log t)$ that are of practical relevance, so that a contour line through $(1/T_a, \log t_a)$ has positive slope $B_a = (\log t_a - \log t_c)/(T_a^{-1} - T_c^{-1})$ which increases both with T_a and with t_a. The coefficients c_0 and c_1 need to be such that the right-hand side of (9.14) is negative so that r is less than 1. The parameter $\mu_{\max}$ represents a notional upper bound for the mean track length before any annealing has taken place and λ allows for a power transformation that extends the class of models.

Figure 9.3 shows some contours of constant mean length for a fanning model that has been fitted to the Durango apatite data in Table 9.1. These contours increase in slope and become closer and more nearly parallel as μ decreases. The points on this figure show the temperatures and times that were used in

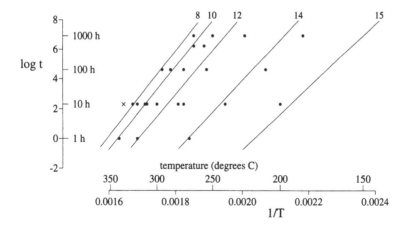

Figure 9.3 *Arrhenius plot showing contours of constant mean length (8, 10, 12, 14 and 15 microns) for a fanning model (9.13) fitted to the Durango apatite data in Table 9.1. The points show the temperatures and times used in these experiments. The cross identifies the (10 h, 335°C) experiment where no tracks were seen.*

the experiments. They have been quite well chosen, though more experiments would be desirable at some lower temperatures and longer times.

There is scope for considering other model equations. Important criteria for models is that they should fit observed data well and also be adaptable to geological time scales. Simplicity, parsimony (i.e., not too many parameters) and interpretability are also desirable.

9.6 Fitting annealing models

In earlier studies, models were fitted to transformed data z_i of the form

$$z_i = \frac{\left(\frac{1-(y_i/y_0)^b}{b}\right)^a - 1}{a}$$

for some empirically determined constants a and b, where y_i and y_0 are the observed mean confined track lengths for experiment i and for the control sample, respectively. The special case $a = 0$ and $b = 1/\lambda$ corresponds to

$$z_i = \log\left(1 - (y_i/y_0)^{1/\lambda}\right),$$

so equation (9.14) might be fitted by non-linear regression to these z_i s.

However, this approach is unsatisfactory for several reasons. These include problems associated with Box-Cox style data transformations, including difficulties of interpretation of parameters, of specifying variance components between and within experiments, and of generalisation to field samples. There are also identifiability problems for some values of a and b. In addition, the use of y_i/y_0 complicates the error structure by introducing both bias and correlation between observations. Furthermore it has been shown that some annealing of freshly formed tracks takes place even at low temperatures (Donelick et al., 1990). Therefore, the mean length for the control sample, even if it were known exactly, is not necessarily the correct normalising quantity

That approach has therefore been superseded by a much simpler one in which the observed mean length y_i is the primary response variable and equations are specified for the expected value and error variance of y_i. For example, for a fanning model, y_i would have expected value μ_i given by (9.13) for the temperature T_i and time t_i determined in experiment i. In general, the error variance will increase with the amount of annealing and may conveniently be specified as a function of μ_i. Laslett and Galbraith (2000) used

$$\text{var}(y_i) \;=\; \sigma^2 \left(1 - \frac{\mu_i}{\mu_{\max}}\right)^{\kappa} + \frac{v_w(\mu_i)}{n_i}, \tag{9.15}$$

where σ and κ were further parameters to be estimated and $v_w(\mu)$ was a known function of μ. The particular form used was

$$v_w(\mu) = \exp\left\{ 2\alpha + 2\beta \left[(\log \mu - \delta) - \sqrt{(\log \mu - \delta)^2 + \gamma}\, \right] \right\},$$

where α, β, γ and δ were determined by fitting the curve $\log v_w(\mu)$ as a function of $\log \mu$ to the log-transformed observed means and standard deviations ($\log y_i, \log s_i$). The first term in equation (9.15) represents variation in mean track lengths "between" experiments — i.e., for aliquots heated at the same T and t on different occasions and etched separately — while the second term represents sampling variation "within" experiment i and which depends on the number n_i of tracks measured. The expected mean length for the control sample was assumed to be less than $\mu_{\max}$ and a model was also specified for the probability of seeing no tracks as a function of μ. This latter was of the form

$$p(\mu) = \frac{e^{b(\mu-a)}}{1 + e^{b(\mu-a)}}$$

for suitably chosen values of a and b, based on past experience and knowledge.

In view of the features seen in Chapter 8, it seems preferable to use, as the response variable, the mean lengths *adjusted* for orientation, rather than the observed (unweighted) mean lengths. These would be found by estimating $E[l|\phi]$ from the length and angle measurements for each annealing run and calculating an appropriate weighted average with respect to the same distribution of ϕ. Uniform weights would give an estimate of μ_{hc}, the theoretical mean length of horizontal tracks; weights proportional to $\sin \phi$ would give an estimate of μ, the mean length over all orientations; and weights proportional to $\sin^2 \phi$ would give an estimate of the equivalent isotropic length. This last choice was suggested by Ravenhurst et al. (2003). Yet another possibility, proposed by Donelick et al. (1999), is the estimated mean length in the direction of the c-axis — that is, the estimate of μ_c in Section 7.11. This is not as sensitive to annealing as the other choices (see Figure 7.8, for example) and is

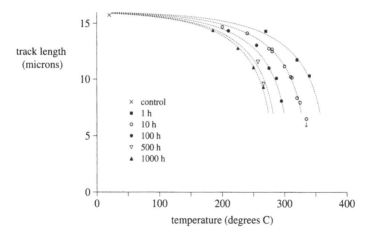

Figure 9.4 *Adjusted mean lengths y_{ai} plotted against temperature for the Durango apatite experiments in Table 9.1. The dotted curves show fitted values from equation (9.13) for each annealing time. A symbol at 7.5 μm with a down arrow represents an experiment in which no tracks were observed.*

therefore probably less satisfactory. All of these require both track lengths and angles to be measured. A more general approach would be to specify a model equation explicitly for $E[l|\phi]$, rather than μ, as a function of temperature and time.

For illustration, some adjusted mean lengths y_{ai} are shown in Table 9.1 for the Durango apatite. These use weights proportional to $\sin \phi$ and are given by

$$y_{ai} = \int_0^{\frac{\pi}{2}} \sin \phi \, \hat{\mu}(\phi) \, d\phi = \hat{\eta} + \tfrac{1}{2}\hat{\xi}$$

using the cosine model in Section 7.11. The equation $\mu(\phi) = \eta + \xi \cos \phi$ was fitted to the length and angle data for each experiment and the residual standard deviation s_{ai} from this fit is also given in Table 9.1. In most cases y_{ai} is less than y_i and more so for the more heavily annealed samples.

Figure 9.4 shows a plot of the adjusted mean lengths against temperatures for the Durango apatite data. This figure shows a pattern typical of laboratory annealing data. As temperature increases, track shortening becomes more rapid and, after sufficient shortening, only a small increase in temperature is needed for the mean length to decrease to a level where no tracks will be seen. For the 10 h, 350°C experiment where no tracks were seen, a point is plotted at a mean length of 7.5 μm along with a down arrow, to indicate that the true mean length probably lies beneath this point.

The dotted lines in Figure 9.4 show fitted values for a fanning model. Fitting was done numerically using maximum likelihood assuming that y_{ai} is normally distributed with mean μ_i of the form in (9.13) and variance of the form in

(9.15) with $\kappa = 1$. Because the point $(1/T_c, \log t_c)$ is a long way from the data region, numerical optimisation can be unstable if these parameters are used directly. It is therefore wise to re-parametrise the model and express c_0, c_1, $\log t_c$ and $1/T_c$ in terms of four other parameters μ_a, μ_b, B_a and B_b corresponding to mean lengths and slopes of contour lines passing through two chosen points $(1/T_a, \log t_a)$ and $(1/T_b, \log t_b)$ on the Arrhenius plot. For well-spaced points within the data region the likelihood function in terms of the latter parameters is easily maximised. The parameter estimates are $\hat{\mu}_{\max} = 16.14$, $\hat{\lambda} = 0.34$, $\hat{c}_0 = -5.90$, $1000\hat{c}_1 = 0.202$, $\log \hat{t}_c = -35.73$, $1000/\hat{T}_c = 0.342$ and $\sigma = 0.43$. Analysis of the standardised residuals suggests that the model fits well.

9.7 Calculating the length distribution from a thermal history

An annealing model derived in the laboratory gives equations for the amount of annealing of tracks held at a constant temperature for a given time. The amount of annealing is usually expressed in terms of the length reduction factor $r = \mu/\mu_{\max}$, c.f., equations (9.13) and (9.14). But tracks in nature have typically experienced varying temperatures $T(t)$ at different times t, such as in Figure 9.1. In that figure and in this section t denotes a point in time, rather than an elapsed time as in equation (9.13). Figure 9.1 uses the standard convention that t represents the time before the present, with the present time denoted by $t = 0$ at the right-hand end of the time axis and the time of deposition t_0 at the left-hand end. Also temperature increases down the vertical scale, as does depth.

Furthermore, tracks are created effectively uniformly over time. Thus, for each time τ from t_0 to the present day, the expected length reduction of tracks formed at time τ due to the thermal history that these tracks experience is calculated (see below). These length reductions are then multiplied by $\mu_{\max}$ and combined into a frequency distribution of lengths, possibly after accounting for variation between individual tracks having the same thermal history. Figure 1.3 shows examples of such frequency distributions.

For tracks formed at time τ, adapting an annealing model to apply to a variable thermal history is usually based on a "lack of memory" assumption proposed by Goswami et al. (1984). Thus, if at time t a track has been annealed to a length reduction r, then it is assumed that any further annealing after t may depend on r and on the temperature at t but not on the specific time-temperature path prior to t. This is akin to the strong Markov property in stochastic processes, though here we are treating the annealing process as being deterministic.

The expected length reduction of tracks formed at time τ may be calculated quite simply by the method of "equivalent time". The thermal history $T(t)$ for $\tau \geq t \geq 0$ is approximated by a step function where the time interval from τ to 0 is broken into n intervals in each of which the temperature is constant. This is illustrated in Table 9.2, where the ith interval goes from time t_{i-1} to

Table 9.2 *Length reduction factor r for equivalent time calculation.*

Time before present	τ	t_1	t_2	t_3	$\cdots$	t_{n-1}	$t_n = 0$
Duration of step		Δ_1	Δ_2	Δ_3	$\cdots$		Δ_n
Temperature during step		T_1	T_2	T_3	$\cdots$		T_n
Total track length reduction	r_0	r_1	r_2	r_3	$\cdots$	r_{n-1}	$r_n = r_\tau$

t_i and is of length $\Delta_i = t_{i-1} - t_i$. During this interval, the temperature is held constant at T_i. If n and the t_i are well chosen then such a step function should closely approximate a given thermal history. There is no need to use the same choices for each τ. The total time from τ to t_i is $\Delta_1 + \Delta_2 + \cdots + \Delta_i$, and the total length reduction in this time is r_i. We wish to calculate $r_n = r_\tau$, the expected length reduction at the present day of tracks formed at time τ.

The calculation is easily illustrated using the parallel Arrhenius model. Let Δ denote an elapsed time for which tracks have been held at a fixed temperature T, say. Then equation (9.11) may be re-arranged to

$$C(r) = \Delta e^{-b/T} \quad \text{and also to} \quad \Delta = C(r)e^{b/T} ,$$

where

$$C(r) = \exp\left\{ \frac{\log(1 - r^{1/\lambda}) - c_0}{c_1} \right\} . \tag{9.16}$$

The reduction r_1 at time t_1 (after an elapsed time Δ_1) is given by

$$C(r_1) = \Delta_1 e^{-b/T_1} .$$

The equivalent time for reduction to r_1 at temperature T_2 is defined to be

$$\Delta_1^{eq} = C(r_1)e^{b/T_2} .$$

From the lack of memory property, the total reduction at time t_2, after an elapsed time $\Delta_1 + \Delta_2$, is the same as if the temperature was fixed at T_2 for time $\Delta_1^{eq} + \Delta_2$. This is given by

$$
\begin{aligned}
C(r_2) &= (\Delta_1^{eq} + \Delta_2)e^{-b/T_2} \\
&= C(r_1) + \Delta_2 e^{-b/T_2}
\end{aligned}
$$

after substituting for Δ_1^{eq}. By the same argument

$$C(r_3) = C(r_2) + \Delta_3 e^{-b/T_3}$$

and after n steps

$$C(r_n) = C(r_\tau) = \Delta_1 e^{-b/T_1} + \Delta_2 e^{-b/T_2} + \cdots + \Delta_n e^{-b/T_n} . \tag{9.17}$$

The length reduction r_τ from the whole thermal history may then be obtained by inverting equation (9.16) to give

$$r_\tau = \left(1 - \exp\left\{ c_0 + c_1 \log C(r_\tau) \right\} \right)^{\lambda} . \tag{9.18}$$

For the parallel model, the amount of annealing for a continuous thermal history $T(t)$ from τ to 0, of tracks formed at τ, can be expressed as

$$C(r_\tau) = \int_0^\tau e^{-b/T(t)}\, dt\,, \qquad (9.19)$$

which is a continuous version of equation (9.17).

The same method may be applied to any annealing model. For the fanning model, equation (9.13) may be re-arranged to

$$C(r) = (\Delta/t_c)^R \quad \text{and also to} \quad \Delta = t_c[C(r)]^{1/R}\,,$$

where $1/R = 1/T - 1/T_c$ and $C(r)$ is given by (9.16) with the appropriate coefficients c_0 and c_1 from the fanning model. The reduction r_1 at time t_1 is given by

$$C(r_1) = (\Delta_1/t_c)^{R_1}\,.$$

The equivalent time for reduction to r_1 at temperature T_2 is

$$\Delta_1^{\text{eq}} = t_c[C(r_1)]^{1/R_2}$$

and the total reduction at time t_2, i.e., after an elapsed time $\Delta_1 + \Delta_2$, is

$$
\begin{aligned}
C(r_2) &= \left((\Delta_1^{\text{eq}} + \Delta_2)/t_c\right)^{R_2} \\
&= \left([C(r_1)]^{1/R_2} + \Delta_2/t_c\right)^{R_2}
\end{aligned}
$$

after substituting for Δ_1^{eq}. By the same argument

$$C(r_3) = \left([C(r_2)]^{1/R_3} + \Delta_3/t_c\right)^{R_3}$$

and so on. In this case it is not so simple to express $C(r_i)$ explicitly in terms of the previous temperatures and times, but $C(r_1)$, $C(r_2)$, ..., $C(r_n)$ can easily be calculated recursively from the above formulae. Hence the length reduction factor r_τ for the whole thermal history may be found from (9.18).

9.8 Inferring times and temperatures from lengths

An observed track length distribution is, at least partly, a consequence of a given thermal history. Of course we would like to invert this relationship to express the thermal history directly in terms of length parameters.

For example, it is tempting to start with an integral equation such as (9.19) and try to invert it to obtain the function $T(t)$ in terms of $C(r)$. Such an approach is impractical for several reasons. High among these is the nature of the annealing relation itself, as illustrated in Figure 9.2. This implies that quite different thermal histories can produce similar length distributions. A second reason is that equations such as (9.19) are based on a deterministic model — as if every track started with exactly the same length and shortened by exactly the same amount after the same thermal history. In reality it is much more likely that the annealing process is stochastic, with variation between tracks, and not invertible. A third reason, not to be discounted, lies in the variability of the observational processes discussed in Chapters 7 and 8.

Therefore, inferences are usually made by "forward modelling" where a schematic thermal history is specified in terms of a small number of temperatures and times — for example, line segments joining two or three points on the (T, t) path. Then parameters of the length distribution can be calculated in terms of these, and hence so can the likelihood function, as illustrated in Section 9.2. This is essentially the approach taken by Laslett (personal communication), Gallagher (1995) and others.

In practice, fission track data by themselves are usually not sufficiently informative to enable a geologist to recover a full thermal history in detail. Such information needs to be supported by other geological knowledge and by other thermal history tools, such as vitrinite reflectance and the uranium-thorium-helium method. Together these can be used to estimate the broad pattern of the thermal history, and possibly some of the finer detail. For example, a geologist, on the basis of accumulated knowledge from other sources, may have a view about the likely evolution of a sedimentary basin. This implies a likely thermal history. Fission track data can be used very effectively to test the plausibility of this scenario, to indicate the direction in which it might need to be modified and to quantify the main events.

9.9 Multi-compositional annealing models

Natural apatites occur in a variety of chemical compositions: fluorapatites, chlorapatites, hydroxyapatites and so on. Many contain some chlorine, which replaces the fluorine in the crystal lattice. For example, Durango apatite has about 0.4% chlorine. Probing of natural apatites has found that grains extracted from a sample of rock tend to have a distribution of chlorine contents. This distribution differs from sample to sample, with most grains having rather less than 0.4% and only the occasional grain having more than 1% chlorine.

It transpires that chlorine content has a considerable impact on the annealing behaviour of fission tracks in apatite, where higher chlorine content implies slower annealing. Comparison of the Durango and Deep River Valley data in Table 9.1 illustrates this. For annealing over a given time and temperature, it is thought that chlorine content is the dominant controlling factor, although various studies have indicated that other less well-defined factors may also play a role.

Such variation between annealing properties of different apatites means that annealing models need to be adjusted to apply to different chlorine compositions. This complicates the relation between track lengths and thermal histories, but at the same time can increase the precision of a given thermal history estimate. Furthermore, because different grains are sensitive to different temperatures, a wider range of thermal histories then can be reconstructed from fission track observations.

9.10 Bibliographic notes

Section 9.2 is based on Galbraith *et al.* (1990, §6). Gallagher (1995) used a joint likelihood of counts and lengths for fitting thermal history models.

Some early discussion of forms of annealing models can be found in Green *et al.* (1985, 1988), Laslett *et al.* (1987) and Duddy *et al.* (1988). Parallel and fanning Arrhenius models (Section 9.5) were fitted to confined track length annealing data by Laslett *et al.* (1987) with further statistical developments in Laslett and Galbraith (1996b), Galbraith and Laslett (1997) and Laslett and Galbraith (2000).

The argument in Section 9.7 follows that in Duddy *et al.* (1988) in a slightly different notation. The ideas in Section 9.8 and elsewhere in this chapter reflect some years of practical experience in commercial applications (Geoff Laslett, personal communication).

Notes on statistical methods

This appendix gives brief notes on some of the statistical concepts used in this book. There are many good text books on probability and statistics, though the vast majority are aimed at students rather than at scientists or practitioners. Well-known books in the latter category include Armitage, Berry and Matthews (2002), Box, Hunter and Hunter (1978) and Snedecor and Cochran (1980). The first of these has recently been updated and published in its fourth edition. While aimed primarily at medical researchers, it is also an excellent general source for both concepts and methods. Ross (1997) is a good text on probability, while Rice (1995) is a standard text on statistical theory and methods. For the reader unfamiliar with a particular topic, a useful source is the *Encyclopedia of Biostatistics* edited by Armitage and Colton and published by Wiley. The second edition (2005) is available both in print and online. The entries are mostly aimed at practitioners or researchers rather than at professional statisticians. Another useful source is the *Encyclopedia of Statistical Sciences* — multiple volumes and updates, edited by Kotz and others and published by Wiley from 1982 on. Readers will no doubt have their own favourite books on statistical methods. These notes aim to explain some of the underlying concepts.

A.1 Poisson processes in one, two and three dimensions

Poisson processes are stochastic models that describe random locations of points along a line, in a plane or in space. They are basic to probability and statistical inference.

In one dimension the Poisson process is usually introduced as describing point events that occur randomly in time — in the sense that, regardless of how many events have happened up to time t and of when they happened, the probability that the next event occurs in a short time interval between t and $t + h$ is the same. For a *homogeneous* process, this probability also does not depend on t and is formally given by $\lambda h + o(h)$ where λ is a constant, called the *rate* of the process, and $o(h)$ is a quantity of smaller order of magnitude than h, i.e., $o(h)/h \to 0$ as $h \to 0$.

Intuitively, the Poisson process is best understood as a limiting case, as $n \to \infty$, of a sequence of Bernoulli trials, with n trials per unit time, each having probability λ/n of success, independent of the outcomes of all other trials. When n is large, the probability of success is small, and the vast majority of trials will result in a failure. The points in time at which the successes occur are effectively points in a Poisson process with rate λ per unit time.

Most properties of Poisson processes can be seen in terms of this concept. For example, the number of successes in a fixed number nt of trials has a binomial distribution with index nt and success probability λ/n. For large n this is effectively a Poisson distribution with mean $nt \times \lambda/n = \lambda t$, which is the expected number of events in a time interval of length t. Numbers of events in two or more non-overlapping intervals have independent Poisson distributions, because they involve different trials. Also, the sum of such numbers must have a Poisson distribution with mean equal to the sum of the separate means, because the sum is just the number of successes in a larger number of trials.

Here are four realisations, over a unit time interval, of a Poisson process with rate $\lambda = 5$ events per unit time.

The locations of the events along the line are indicated by crosses. The number of events in each is drawn from a Poisson distribution with mean 5. If we concatenated the four intervals we would have a realisation over four time units. If we superimposed them, we would have a realisation over one time unit of a Poisson process with rate $\lambda = 20$ events per unit time.

Another property of the homogeneous Poisson process is that the time intervals between successive events are independently drawn from an Exponential distribution with rate parameter λ. This provides a simple way of simulating realisations of the process: draw a random value from the exponential distribution to determine the time of the first event, then draw another value and add it to the first to get the time of the second event, and so on.

Yet another property, that also provides a simple simulation method, is: given that n events occurred in an interval from s to $s + t$, the *times* at which they occurred are a random sample of n observations from a uniform distribution over that interval. The above realisations were simulated by first drawing a random value, n say, from a Poisson distribution with mean 5, and then drawing n values independently from a uniform distribution. For the first realisation, $n = 4$ was obtained, for the second, $n = 2$, and so on. ·

One consequence of this property, that we use repeatedly, is: suppose that n events have happened in a time interval of length t, then the number of these that are located in a sub-interval of length s must have a binomial distribution with index n and parameter s/t. This is because each of the n points independently has probability s/t of occurring in the sub-interval. Another aspect of this is that, if we have two independent counts from Poisson distributions with means μ_1 and μ_2, then the conditional probability distribution of the first of them, given that their sum equals n, is binomial with index n and parameter $\mu_1/(\mu_1 + \mu_2)$. Intuitively, this distribution gives the relative frequencies of the various possible values of the first count amongst all pairs of counts that add to n.

In this book we are concerned with Poisson processes in two and three dimensions, the former to describe the spatial locations of the points of intersection of fission tracks with a plane, and the latter to describe locations of fissioned uranium atoms within a crystal. The last concept above gives the easiest description of Poisson processes in higher dimensions. Thus, for a spatial Poisson process, the number of points within any plane region has a Poisson distribution, and numbers in disjoint regions are independent. Furthermore, given that n points occur in a particular region, their locations are randomly drawn from a uniform distribution over it. If the region has area A then the Poisson distribution has mean $A\rho$, where ρ is the expected number of points per unit area and is often called the *density* in this context.

The figure below shows four simulated realisations of a spatial Poisson process with $\rho = 25$ over unit square regions. Each was simulated by first drawing n from a Poisson distribution with mean 25, and then drawing n independent x and y coordinates from uniform distributions between 0 and 1. The visual appearance, with some areas containing several points close together and other quite large empty areas, is typical of a spatial Poisson process. Another way to envisage this process is to imagine a square grid of a large number of points, say 1000×1000, and for each point in turn either delete it with probability $1 - p$ or retain it with probability p, where in the present case $p = 25 \times 10^{-6}$. The crosses then indicate the locations of the retained points.

Exactly the same principles apply to Poisson processes in three dimensions. Numbers of points in disjoint regions have independent Poisson distributions,

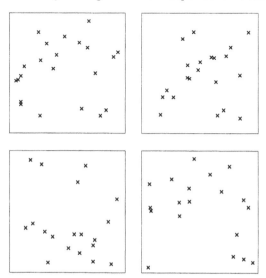

and given n points in a region, their locations are independently drawn from a uniform distribution over it. Again, this can be envisaged in terms of a three-dimensional grid of points, each of which is independently retained with a small probability. In fact it is not necessary for the initial arrangement of points to be a regular grid — a realisation of any homogeneous point process will do. This is the essence of the concept of "independent thinning" of a point process that in the limit becomes a Poisson process, and is the reason why locations of fissioned uranium atoms form a Poisson process, as discussed in Chapter 2.

A.2 Notes on the Poisson distribution

The Poisson distribution is a probability model for counts. Consider a count N that is a random quantity with expected value $A\rho$, where A is known (e.g., A is the area of grain surface) and ρ is an unknown density to be estimated. If N has a Poisson distribution then the probability of observing the value $N = r$, for $r = 0, 1, 2, \ldots$, is given by

$$P(N = r) = \frac{(A\rho)^r e^{-A\rho}}{r!}. \tag{A.1}$$

This distribution has mean $A\rho$ and variance $A\rho$, equal to the mean. Its standard deviation is therefore $\sqrt{A\rho}$ and its coefficient of variation is $\sqrt{A\rho}/A\rho = 1/\sqrt{A\rho}$, which is the reciprocal of the square root of the mean. There are other probability models for counts also. The Poisson distribution is applicable when the counts effectively arise from an underlying Poisson process as in Section A.1.

Suppose we observe $N = r$ and we wish to estimate ρ. The natural estimate is the observed number tracks per unit area, that is,

$$\hat{\rho} = \frac{r}{A}. \tag{A.2}$$

The standard error of this estimate is

$$se(\hat{\rho}) = \sqrt{\frac{\rho}{A}} \tag{A.3}$$

and the relative standard error is

$$\frac{se(\hat{\rho})}{\rho} = \frac{1}{\sqrt{A\rho}}. \tag{A.4}$$

This depends on the unknown ρ, so in order to *estimate* the relative standard error it is usual practice to substitute $\hat{\rho}$ into (A.4) to give

$$\frac{se(\hat{\rho})}{\rho} \approx \frac{1}{\sqrt{A\hat{\rho}}} = \frac{1}{\sqrt{r}}. \tag{A.5}$$

That is, the relative standard error of a mean or density estimated from a Poisson count is approximately the reciprocal of the square root of the count. Likewise, the standard error is approximately the square root of the count.

When the observed count r is small, the approximation (A.5) will be poor, and when $r = 0$ it breaks down completely. Furthermore when the true mean $A\rho$ is small the standard error or relative standard error, whether known or estimated, is not a very good measure of precision, in the sense that the interval $\hat{\rho} \pm 2\text{se}(\hat{\rho})$ does not provide an approximate 95% confidence interval for ρ. There are other methods, and indeed statistical tables, for calculating confidence limits for ρ from small counts (e.g., Pearson and Hartley, 1966, Table 40).

The case $r = 0$ sometimes causes confusion. Formally the *estimate* of ρ is then zero, but this does not mean that ρ is zero. It is quite possible to observe a zero count when ρ is greater than zero — in fact this is likely when ρ is small. This example highlights the need to distinguish between the true value of a parameter and an estimate of it. Statisticians usually note this distinction by putting a hat on an estimate — thus $\hat{\rho}$ denotes an estimate of ρ.

As well as being the "obvious" estimate, $\hat{\rho}$ given by (A.2) is also the *maximum likelihood* estimate of ρ. It is the value of ρ for which $P(N = r)$, given by (A.1), is a maximum. In other words it is the value of ρ for which the observed value of N is most probable. In statistical theory, maximum likelihood estimates are known to be efficient (under general conditions), which means, roughly, that they make full use of the available data. When $r = 0$, $\hat{\rho} = 0$ is still the maximum likelihood estimate, but the general theory is not automatically applicable because the maximum of the likelihood occurs when ρ is at the boundary of its possible values (see Section A.10).

Now consider several counts. If two independent counts N_1 and N_2 have Poisson distributions with means $A_1\rho_1$ and $A_2\rho_2$, then their sum $N_1 + N_2$ has a Poisson distribution with mean $A_1\rho_1 + A_2\rho_2$. This property extends to sums of any number of independent counts.

Suppose we have a count over an area in each of several grains which have the same unknown density ρ. If the total count over all grains is N and the total area is A, then N has a Poisson distribution with mean $A\rho$, where A is the total area. So the maximum likelihood estimate of ρ is still given by (A.2) with relative standard error given by (A.4).

A.3 Relation between the binomial and Poisson distributions

The binomial distribution with index n and parameter θ is a probability model for a count N that is restricted to the values $0, 1, \ldots, n$. It represents the probability distribution of the number N of "successes" in n independent "trials", where each trial has the same chance θ of a success. The parameter θ often represents an unknown probability or proportion that we wish to estimate. The probability that $N = r$, for $r = 0, 1, \ldots, n$, is given by

$$P(N = r) \;=\; \binom{n}{r} \theta^r (1 - \theta)^{n-r}. \tag{A.6}$$

This distribution has mean $n\theta$ and variance $n\theta(1-\theta)$. Since $0 < \theta < 1$, the variance is less than the mean (ignoring the deterministic cases when $\theta = 0$ or 1). The Poisson distribution is often introduced as a limiting case of the binomial. If we put $\theta = A\rho/n$ in (A.6) and let $n \to \infty$, then $P(N = r)$ tends to the value given by (A.1). Thus if the number of trials n is large and the probability of success θ is small, the number of successes has (approximately) a Poisson distribution with mean $n\theta$.

There is another relation between the Poisson and binomial distributions, referred to in Section A.1. Suppose that two independent counts N_s and N_i have Poisson distributions with means $A\rho_s$ and $A\rho_i$, and we observe their sum $N_s + N_i = m$, say. Then the conditional probability that $N_s = r$ given that $N_s + N_i = m$ is given by $P(N = r)$ in (A.6), with $n = m$ and

$$\theta \;=\; \frac{A\rho_s}{A\rho_s + A\rho_i} \;=\; \frac{\rho_s}{\rho_s + \rho_i} \;=\; \frac{\rho_s/\rho_i}{1 + \rho_s/\rho_i}. \tag{A.7}$$

The right-hand form of (A.7) shows that θ depends just on the ratio ρ_s/ρ_i and not on the separate densities. In fission track dating it is this ratio that we wish to estimate (see Chapter 3). We may write it as

$$\frac{\rho_s}{\rho_i} \;=\; \frac{\theta}{1-\theta}. \tag{A.8}$$

Treating $N_s + N_i$ as being fixed, the standard estimate of θ is $\hat{\theta} = N_s/(N_s + N_i)$ and so the estimate of ρ_s/ρ_i is N_s/N_i. This is the usual estimate, but we have now found it without estimating ρ_s and ρ_i separately.

As it turns out, because the uranium content of the grain is unknown, the total count does not tell us anything about θ, and we are led to the same answer as if we had estimated the uranium content (see Section 5.1). This use of the binomial distribution is convenient in more complicated situations. For example, it underlies many formulae applicable to the external detector method, including (3.15), (3.16), (3.20), (3.22) and (3.23) in Chapter 3 as well as many of the models analysed in Chapters 5 and 6.

A.4 Standard errors and confidence intervals

Suppose that an experiment produces an estimate $\hat{\rho}$ of an unknown parameter ρ. The *standard error* of $\hat{\rho}$, denoted by se$(\hat{\rho})$, is a measure of the precision with which ρ has been estimated, where a small standard error indicates high precision. It is the standard deviation of all of the possible estimates that the experiment might (hypothetically) have produced. The theoretical probability distribution of all possible estimates is called the *sampling distribution* of $\hat{\rho}$.

For example, imagine repeatedly drawing a random value r from a Poisson distribution with mean $A\rho$ and for each r calculating $\hat{\rho}$ by equation (A.2). Then the frequency distribution of estimates so obtained is the sampling distribution $\hat{\rho}$. The standard error of an estimate is the standard deviation of its sampling distribution. In the present example, a mathematical formula for the standard error of $\hat{\rho}$ can be derived and is given by (A.3). Nearly always in practice the true standard error is not known and must itself be estimated. Some statistics textbooks use a different notation, such as ese($\hat{\rho}$), to denote an estimated standard error, but standard errors quoted in applications are usually implicitly assumed to be "estimated" or "approximate" without any notational distinction.

Often the sampling distribution of $\hat{\rho}$ is approximately normal or Gaussian. Then $\hat{\rho} \pm 2\,\mathrm{se}(\hat{\rho})$ is an approximate 95% *confidence interval* for ρ. This is sometimes called a "two-sigma" interval. Such an interval has approximately a 0.95 chance of including the true value of ρ, in the sense that it will do so in about 95% of (hypothetical) experiments. A useful interpretation is to regard a 95% confidence interval as a range of values of ρ that are reasonably consistent with the observed data. This is based on the idea of inverting a series of significance tests. For each possible value of ρ we imagine testing the null hypothesis that that is the true value. Then the set of values of ρ for which the p-value is greater than 0.05 constitutes a 95% confidence interval.

Sometimes the sampling distribution is not approximately normal. Then it can help to transform the parameter to a new scale. For example one might use $\beta = \log \rho$, estimate β and its standard error, and then transform estimates and confidence intervals back to the original scale to give a confidence interval for ρ of $\exp\{\hat{\beta} \pm 2\,\mathrm{se}(\hat{\beta})\}$. This approach often gives better results for fission track age estimates, especially those based on small counts.

The *relative standard error* of an estimate $\hat{\rho}$ is defined as $\mathrm{se}(\hat{\rho})/\rho$. It is the standard error as a fraction of the true value. It is often multiplied by 100 and quoted as a percentage. Because it measures precision relative to the size of the quantity of interest, it can be more useful than the absolute standard error for comparing estimates of quantities of different sizes. Of course this concept is not sensible for quantities such as differences in means whose "size" might be zero or negative. Provided that it is not too large, the relative standard error is closely approximated by the standard error of the natural logarithm of the estimate, that is,

$$\frac{\mathrm{se}(\hat{\rho})}{\rho} \approx \mathrm{se}(\log \hat{\rho}). \qquad (A.9)$$

This is a useful principle. The approximation is rough for large relative errors, but is very good below about 0.5 (i.e., a 50% relative standard error or less).

In fission track dating it is the relative standard error of the age estimate that directly reflects the precision of the experiment. By contrast, the absolute standard error also depends on the grain's age — other things being equal, an age estimate for an older grain will have a larger standard error simply because the grain is older. Now estimating t in equation (3.10) is equivalent to estimating η, where

$$\eta = \log\left(\tfrac{1}{2}\zeta\rho_d\frac{\rho_s}{\rho_i}\right) = \log\tfrac{1}{2} + \log\zeta + \log\rho_d + \log\rho_s - \log\rho_i \qquad (A.10)$$

and then substituting this in the equation

$$t = \tfrac{1}{\lambda} \log (1 + \lambda e^{\eta}). \qquad (A.11)$$

Inverting equation (A.11) gives

$$\eta = \log \left(\frac{e^{\lambda t} - 1}{\lambda} \right) \approx \log t \qquad (A.12)$$

so that η is approximately the natural logarithm of the age.

This idea is useful both conceptually and in practice. We obtain an estimate $\hat{\eta}$, its standard error $se(\hat{\eta})$, and two-sigma confidence limits $\hat{\eta} - 2\,se(\hat{\eta})$ and $\hat{\eta} + 2\,se(\hat{\eta})$. Substituting $\hat{\eta}$ into equation (A.11) gives the age estimate $\hat{t}$. Substituting each limit into equation (A.11) gives corresponding confidence limits for t. The relative standard error of $\hat{t}$ is approximately

$$\frac{se(\hat{t})}{t} \approx \frac{1 - e^{-\lambda t}}{\lambda t}\, se(\hat{\eta}). \qquad (A.13)$$

The factor multiplying $se(\hat{\eta})$ in (A.13) is close to 1 — even for t as large as 500 Ma it is 0.962 — so it might as well be neglected, especially as $se(\hat{\eta})$ itself is only approximate. Thus for most practical purposes

$$\frac{se(\hat{t})}{t} \approx se(\hat{\eta}). \qquad (A.14)$$

This result follows immediately from equation (A.12), that is,

$$se(\hat{\eta}) \approx se(\log \hat{t}) \approx se(\hat{t})/t.$$

The concepts of standard errors and confidence intervals apply to any parameter, not just to a mean or a density or a fission track age.

A.5 Components of error

Many explanations of standard errors, including the above one, may give the impression that there is a single true standard error of a given estimate, albeit one that may be estimated in different ways. But in fact, different standard errors are applicable for different purposes.

For example, equation (3.13) accounts for error in estimating ρ_d and ζ as well as ρ_s/ρ_i. This would be applicable for comparison with, say, a stratigraphic age or an independently determined age. But if the estimate was to be compared with a fission track age from a different sample determined by the same analyst, using the same zeta value, then the contribution from error in $\hat{\zeta}$ should be omitted. This is because any error in $\hat{\zeta}$ will affect both estimates in the same way and will cancel out in the comparison. Likewise, when comparing age estimates from grains that were irradiated together, the contribution from error in estimating ρ_d should be omitted.

Usually there are several sources of error that contribute to the precision of an estimate. Sources of variation in the experimental processes should, as much as possible, be identified and their effects estimated independently.

Some natural variation may be reduced by experimental design, but there will usually be a residual error from unknown sources. In fact, standard errors are very rarely absolute. For example, inter-laboratory studies (e.g., Miller *et al.*, 1993) have revealed that there are differences between estimates from different laboratories that are not accounted for in (3.13). So when combining or analysing estimates from different laboratories, further components of error may need to be considered.

A.6 Precision and accuracy

In the physical sciences a useful distinction is made between *precision* and *accuracy*. The former refers to how close an estimate is likely to be to its expected (or average) value while the latter refers to how close this expected value is to the true value of the quantity being estimated. Corresponding statistical concepts are the *variance* and *bias* of an estimator. The standard error of an estimate (the square root of the variance) is a measure of lack of precision while the bias is a measure of lack of accuracy. A closely related concept is the distinction between *random* and *systematic* errors, where the former contribute to lack of precision and the latter to lack of accuracy.

However, these distinctions are not absolute and can depend on the types of comparisons being made. For example, differences between estimates made by different observers might for some purposes be regarded as systematic and for other purposes as random.

A.7 Statistical significance tests and p-values

A statistical significance test is a method of assessing whether observed data agree with a hypothesis. It is a form of uncertain inference.

The method works as follows. One proposes a "null" hypothesis, denoted by H_0, which is an assertion about the "population" from which the data came. From the data, one calculates a test statistic and then assesses its statistical significance by calculating a p-value. The test statistic may be the data itself, but usually it is a summary statistic designed to measure a feature of interest. The p-value is the probability of obtaining a value of the test statistic as extreme as, or more extreme than, the calculated value, assuming that the null hypothesis is true. The smaller the p-value, the stronger is the evidence that the data do not agree with the null hypothesis.

In order to know what "more extreme" means, it is necessary to have specified an alternative hypothesis H_1. Then "more extreme" means "further in a direction that would be less probable if H_0 were true than if H_1 were true". In general, H_0 is a precise statement that determines the sampling distribution of the test statistic, while H_1 may be vague and serves to determine what is meant by "more extreme".

For example, for the chi-square test in Section 3.8, the null hypothesis is that all grains have the same *true* fission track age and the alternative is that some grains have different true ages. The test statistic is χ^2_{stat} given by (3.20). This measures the discrepancy between the observed counts and what they would be expected to be if all grains had the same age. If H_0 is true, then the sampling distribution of χ^2_{stat} is approximately χ^2 with $n-1$ degrees of freedom. If H_1 is true, i.e., if some grains had different true ages, then χ^2_{stat} would tend to be larger. So the p-value is the probability that a random value from the $\chi^2(n-1)$ distribution is greater than the observed value of χ^2_{stat}. The larger χ^2_{stat} is, the smaller the p-value is and the stronger the evidence is that not all grains have the same age.

By convention, a p-value less than 0.01 represents strong evidence against H_0, a p-value between 0.01 and 0.05 represents moderate evidence against H_0, and a p-value greater than 0.05 represents little or no evidence against H_0, but these are not hard and fast rules. A p-value of 0.04 does not represent a very different amount of evidence from one of 0.06, so this should be recognised. A largish p-value, greater than 0.2, say, is often interpreted as "the data are consistent with H_0".

The nature of the inference is indirect — in particular, a p-value does not tell us the probability that H_0 is true. Its interpretation is essentially based on convention. Understanding the following points helps to avoid pitfalls:

• Data that are consistent with H_0 are usually consistent with other hypotheses too. A large p-value — even 0.9 or higher — means lack of evidence against H_0, rather than direct evidence in favour. For example in the chi-square test in Section 3.8, the p-value might be large because the track counts are small, so there is not much evidence in the data. Mistakes often come from assuming something is true just because there is little or no evidence against it.

• Data can show strong *evidence* against H_0, even when H_0 is close to the truth. Even a very small p-value does not necessarily imply that there is a large or *important* departure from H_0. For example, one might obtain a very large χ^2_{stat} and small p-value as a result of having very informative data — large numbers of tracks per grain — but the true ages, though different, might not be very different.

• A significance test is limited in scope; it is only concerned with whether H_0 is true and usually some form of estimation is also wanted. For example, suppose that the chi-square test gave a p-value of less than 0.01. This would indicate evidence that the true fission track ages varied between grains. One might then want to estimate how much they varied — e.g., by estimating the age dispersion or by fitting a mixture model such as in Chapter 5 or 6.

For a general discussion of p-values, see, for example, Armitage *et al.* (2002, p. 88). In the physical sciences, emphasis is usually more on estimation than on significance tests. The latter are often used for checking assumptions and for comparing the goodness of fit of competing models.

A.8 Radial plots

A radial plot is a graphical method for comparing several estimates that have different precisions. For example, we may wish to compare fission track age estimates from a number of single grains, where each estimate has a different standard error or relative standard error. It is inherently not straightforward to do this, and graphs that do not properly account for the varying precisions are likely to be misleading, or at best, difficult to interpret.

A further feature of many applications, including fission track ages, is that not only do the precisions vary, but also they tend to be related to the size of the parameter being estimated — usually with a larger estimate having a larger standard error. Then it is preferable to compare estimates on a transformed scale, where the precisions are (at least approximately) independent of the sizes of the parameters. If the *relative* standard errors are independent of the sizes of the quantities being estimated (i.e., the standard errors are proportional to the sizes of the estimates), then plotting on a log scale will achieve this. This applies to other graphs as well as to radial plots. But, unlike other graphs, radial plots are designed so that estimates can easily be compared while properly taking account of their differing precisions.

Suppose we have a number of estimates $z_1, z_2, \ldots, z_n$ with known standard errors $\sigma_1, \sigma_2, \ldots \sigma_n$. Often z_i will be on a transformed scale, for the reason mentioned above. Then a radial plot is an (x, y) scatter plot of the points

$$x_i = 1/\sigma_i \quad \text{and} \quad y_i = (z_i - z_0)/\sigma_i,$$

where z_0 is a convenient reference value. It is a scatter plot of *standardised* estimates y against *precisions* x, where the precision is defined as the reciprocal of the standard error of z. It is usually best visually to choose z_0 to be a central value so that the y_i scatter around zero. For example, for a sample of single grain fission track age estimates, z_0 could correspond to either the central age or the pooled age for that sample. Sometimes it is useful for z_0 to represent a reference value of interest, such as an independently determined age or a stratigraphic age.

Because the y_i all have unit standard error, the same error scatter along the y scale applies to all points, which can be judged in terms of numbers of standard errors on this axis. A useful design feature is to draw the y-axis to extend just from -2 to $+2$, or otherwise to highlight this range, so that it displays a "two-sigma" error bar that can mentally be placed on any point. Now for any point (x, y), the value of $z - z_0$ is given by ratio y/x, i.e.,

$$\frac{y}{x} = \frac{(z - z_0)/\sigma}{1/\sigma} = z - z_0.$$

This is the *slope* of the line from $(0, 0)$ through (x, y). All points lying on this line have the same value of $z - z_0$ and hence of z. Points further from the origin (i.e., with larger x) have higher precisions. We may display the z values by drawing a scale of slopes, for example along an arc of a circle on the

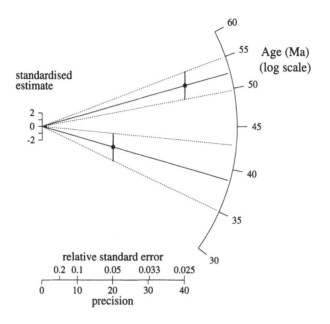

Figure A.1 *Principle of a radial plot with estimates on a log scale.*

right-hand side of the graph. If z is on a transformed scale we may display the estimates either on the transformed scale or on the original scale, or both.

Figure A.1 illustrates these principles for estimates plotted with respect to a log scale. This figure is centered at an age of 45 Ma, so that $z_0 = \log 45 = 3.807$. The age scale is shown, but the log age (z) scale is not shown explicitly. There are two hypothetical points: the first with $z_1 = 3.957$, $\sigma_1 = 0.025$, plotted at

$$x_1 = 1/\sigma_1 = 40, \quad y_1 = (z_1 - z_0)/\sigma_1 = 6$$

and the second with $z_2 = 3.657$, $\sigma_2 = 0.05$, plotted at

$$x_2 = 1/\sigma_2 = 20, \quad y_2 = (z_2 - z_0)/\sigma_2 = -3.$$

For the first point, the line joining $(0,0)$ to $(40,6)$ has slope $6/40 = 0.15 = z_1 - z_0$, which corresponds to an age of $\exp\{z_1\} = \exp\{z_0 + 0.15\} = 52.3$ Ma. Figure A.1 also shows a ± 2 error bar centered at $(40, 6)$, i.e., going from $(40, 4)$ to $(40, 8)$. The dotted radii joining the ends of this error bar give a two-sigma confidence interval for the age going from $\exp\{z_0 + 4/40\} = 49.7$ Ma to $\exp\{z_0 + 8/40\} = 55.0$ Ma. For the second point, plotted at $(20, -3)$, the estimated age is $\exp\{z_0 - 3/20\} = 38.7$ Ma, and corresponding calculations give a two-sigma confidence interval going from $\exp\{z_0 - 5/20\} = 35.1$ Ma to $\exp\{z_0 - 1/20\} = 42.8$ Ma. The same ± 2 error applies to each point but, because the second point has a lower precision, the confidence interval produces a wider range for the age.

Figure A.1 also shows a scale of relative standard errors, based on the relation

$$se(\log \text{age}) \approx \frac{se(\text{age})}{\text{age}} .$$

For the first point, the relative standard error is $\sigma_1 = 0.025$, or 2.5%, and for the second point the relative standard error is $\sigma_2 = 0.05$, or 5%, which can be read off.

A very useful property of the radial plot is that, if there are several estimates of a common true quantity, and their precisions are correctly determined, then the (x, y) points will scatter homoscedastically, with unit standard deviation, about a straight line through $(0, 0)$. The slope of this line indicates the common parameter value. If the quantity being estimated varies, there will be more y scatter at larger x values. Also, if the precisions are not correct the graph should show a corresponding departure from this homoscedastic pattern.

A radial plot of fission track age estimates is shown in Figure 3.2. Each point has unit standard error in the y direction. This means that each point has equal status with respect to variation on the y scale so that comparisons are straightforward. At the same time it is clear that points with larger x values are more informative as they determine z more precisely. For these data, the 20 points scatter homoscedastically within a ± 2 band, indicating that the estimates are consistent with a common value. In Figure 3.3 the same data are plotted but using a different z_0. The points scatter about a different radius, but they are still consistent with a common age.

Radial plots have other useful features. They sort the data so that the less precise grains fall to the left and the more precise grains fall to the right. If the individual estimates are consistent with a common age, about 95% of the points should fall within a band ± 2 units vertically around some radial line. If the data arise from two distinct populations — that is, each grain has one of two different true ages — then the points should mostly lie within two such bands. This can be seen in Figure 5.1 for example. Other specific patterns may also be recognisable. A more complicated, but very informative, pattern can be seen in Figure 6.2. This shows variation in fission track ages between grains within samples, the over-dispersion with respect to Poisson variation, and the variation (in both mean and over-dispersion) between samples. Sometimes *under-dispersion* can be seen, which is also informative (e.g., Galbraith, 1990, Figure 5). A radial plot is therefore both a valid and informative way to display fission track age estimates, which often have substantially differing precisions. Because their precisions are displayed it is possible to compare estimates sensibly, and because they are displayed as points it is possible to put a lot of information on one graph — one of the secrets of good visual displays.

When plotting fission track age estimates, we typically use a transformed scale for z for the reasons given above. For estimates based on moderately large counts, a log transformation is usually satisfactory. A nice feature of this choice is that the relative standard errors can be displayed directly on the x scale as in Figure A.1. For single grain ages with some small counts, though, it is better to use the angular transformation as in equations (3.15) and (3.16). In this case we can display the total numbers of tracks (spontaneous + induced) on the x scale, as in Figure 3.2. In either case it is useful to display an age scale in Ma (albeit not in equal divisions of Ma) rather than the scale of values of z. The statistical basis and other applications of radial plots are described by Galbraith (1988, 1990, 1994).

A.9 Histograms and "probability density" plots

A histogram is a graph of the information in a grouped frequency table derived from observations on a quantitative variable. For each group, a rectangle is drawn with base equal to the group width and area proportional to the frequency for that group. Histograms are good graphs for seeing not only location and spread, but also the *shapes* of distributions, when large enough samples of observations are available. For example they are very useful for looking at fission track lengths, as in Figure 8.2.

Histograms have sometimes been used to plot fission track age estimates obtained by the external detector method. But they are usually poor in this situation because the estimates have different standard errors, many of which are not small compared with any variation there might be in the true ages. A histogram drawn with respect to the age scale (i.e., equal divisions of age) will usually be positively skewed in shape, partly because larger estimates have larger standard errors, so will scatter more. Extreme points may reflect low precision rather than high age. This feature can be improved by using a transformed scale, such as a log scale, to eliminate the relation between the size of an estimate and its precision, but the estimates still have different precisions, which is why a radial plot is needed.

However, single grain ages obtained by the population method *can* be transformed so that their standard errors are approximately equal and independent of the sizes of the estimates. This is done in the right-hand panels of Figures 4.1 and 4.2. A histogram is useful here. It is not quite showing us what we want, but we can make sense of it by comparing it with the same graph constructed from the induced track counts, as in the above-mentioned figures. Such a transformation is not possible for estimates obtained by the external detector method, though.

There is a deeper point here. For simplicity, imagine that the observed estimates are of the form

$$z_i = \beta_i + e_i, \qquad\qquad (A.15)$$

where β_i is the true value and e_i is the error for estimate i. The error e_i is assumed to be drawn from a distribution with mean 0 and known standard deviation σ_i which differs for each i. In fission track analysis, σ_i depends on the measurement procedures used and the physical properties of the ith grain.

What we want is a picture of the distribution of β_i s. We may estimate this by fitting a suitable mixture model, but a histogram of z_i s will not give us this in general. The latter may be informative when the σ_i s are small, but otherwise its shape is largely determined by the e_i s, which are from an arbitrary mixture.

There is a variant of the histogram, sometimes called a *weighted histogram* or a *probability density plot* which I do not recommend. Here the observation z_i is replaced by a Gaussian probability density function centered at z_i and with standard deviation proportional to σ_i. Then these probability density functions are added point-wise to produce a smooth "frequency" curve (e.g., Hurford *et al.*, 1984; Brandon, 1992). The meaning of "frequency" here is not straightforward. The term "weighted histogram" derives from the fact that each frequency is a sum of different fractional contributions from each observation, though each observation still contributes one unit of area under the curve. The construction of such a graph is akin to that of a kernel density estimate, as was noted by Brandon (1996), but with a different kernel bandwidth for each observation. Variants have also been proposed where each observation contributes a different area.

This graph looks intuitively appealing, firstly because it produces a nice smooth curve and secondly because it seems to take account of the differing precisions. Unfortunately, though, it does not do so properly. In fact, it adds extra error variation to the estimates — that is, the variation in e_i contributes twice — and it does not give a true picture of the distribution of β_i s in (A.15) no matter how many estimates there are. Inasmuch as it is a smoothed histogram, this graph can be informative when the variation in the β_i s is large compared with the σ_i s. But the presence of several estimates with low precision can obscure information, even when there are other high precision estimates in the sample. Then the shape of the "probability density" curve is even more dependent on the error distribution than that of a histogram. Some examples of how the method can fail are given in Galbraith (1998).

A.10 Parametric models and likelihood inference

The basic ingredients of parametric statistical inference are *data* and a *probability model* that specifies the probability of the observed data (and of data that might otherwise have arisen) in terms of some unknown *parameters*. For example, as in Section A.2, the data might be a single count r over an area A, with a probability model that r is drawn from a Poisson distribution with mean $A\rho$ where ρ is an unknown parameter.

A.10.1 The likelihood function

The *likelihood function* is formally defined as any function that is proportional to the probability of the data, when considered as a function of the unknown parameters. Nearly always one works with the the natural logarithm of the likelihood function, called the *log likelihood*, which we denote by L. This is the

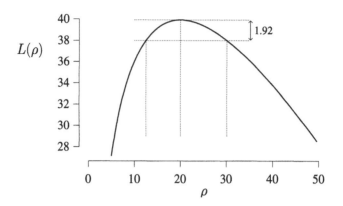

Figure A.2 *Log likelihood function* $L(\rho) = 20\log\rho - \rho$ *for a count* $r = 20$ *from a Poisson distribution with mean* ρ.

log of the probability of the data, plus an arbitrary constant, considered as a function of the unknown parameters. In our simple example

$$\text{prob(data)} = P(N = r) = \frac{(A\rho)^r\, e^{-A\rho}}{r!}$$

and the log likelihood function is the function of ρ given by

$$
\begin{aligned}
L(\rho) &= r\log A + r\log\rho - A\rho - \log r! + \text{constant} \\
        &= r\log\rho - A\rho + \text{constant},
\end{aligned}
\tag{A.16}
$$

where the additive terms $r\log A$ and $-\log r!$ have been incorporated into the constant as they do not depend on ρ. The absolute values of the likelihood or log likelihood have no useful meaning: it is ratios of likelihoods or differences of log likelihoods that are of interest.

The log likelihood function (A.16) is plotted in Figure A.2 for the data $r = 20$ and $A = 1$ and setting the constant equal to zero. It can be seen that L is largest for values of ρ near 20. For other positive values of r, the curve has a similar shape, increasing to a maximum and then decreasing as ρ increases from zero, though it is more skewed for smaller values of r. When $r = 0$ (for example, if no tracks were seen intersecting an area A) the log likelihood decreases linearly, with slope $-A$, from a maximum value at $\rho = 0$. This is simply saying that the larger ρ is, the less likely it is that one would observe $r = 0$, and the more so for larger A. We discuss this figure further in Section A.10.4.

In Section 5.1, the data are matched counts of spontaneous and induced fission tracks for n crystals, denoted by N_{su} and N_{iu} for $u = 1, 2, \ldots, n$. The probability model asserts that all counts have independent Poisson distributions with means given by $\mathrm{E}(N_{su}) = e^{\beta}\mu_u$ and $\mathrm{E}(N_{iu}) = \mu_u$. The unknown parameters are $\mu_1, \mu_2, \ldots, \mu_n$ and β, where the μs are proportional to the amounts of uranium in the relevant parts of the crystals and β is the parameter of interest that relates to the common age of the crystals. Because the Poisson distributions are independent, the probability of the data is the product of the Poisson probabilities for the individual counts, that is,

$$\text{prob(data)} = \prod_{u=1}^{n} \frac{(e^{\beta}\mu_u)^{N_{su}} e^{-e^{\beta}\mu_u}}{(N_{su})!} \frac{(\mu_u)^{N_{iu}} e^{-\mu_u}}{(N_{iu})!}$$

and the log likelihood function is

$$L = \log\big(\text{prob(data)}\big) + \text{constant}$$

$$= \sum_{u=1}^{n} \Big\{ N_{su}\beta - e^{\beta}\mu_u + (N_{su} + N_{iu})\log\mu_u - \mu_u \Big\} + \text{constant}$$

after re-arranging. This is regarded as a function of $\mu_1, \mu_2, \ldots, \mu_n$ and β. Again the sums of the logarithms of the factorials of the counts, which do not depend on any unknown parameters, have been added to the constant.

Sometimes a probability model is specified in terms of a continuous distribution, such as a Normal distribution or an Exponential distribution. In such cases many text books define the likelihood function to be proportional to the joint probability *density* of the distribution of the observations, evaluated at the data values. This can create technical problems and strictly speaking is not correct. This is because observed data are *discrete*, in the sense that, although they may come from a continuous distribution, observations are measured (or recorded) to a given accuracy.

For example, suppose the data are values of X, measured to two decimal places, and the model is that X has a Normal distribution with unknown mean μ and standard deviation σ. Suppose we observe the value 10.22. Then we have not really observed the event $X = 10.22$ but rather the event $10.215 \leq X < 10.225$. The probability of this datum is

$$\text{prob}(10.215 \leq X < 10.225) = \int_{10.215}^{10.225} f(x)\,dx \approx 0.01f(10.22),$$

where $f(x)$ is the probability density function of the Normal distribution, i.e.,

$$f(x) = \frac{1}{\sqrt{2\pi}} \exp\left\{ \frac{1}{2}\left(\frac{x-\mu}{\sigma}\right)^2 \right\}.$$

The integral is the area under the Normal curve between 10.215 and 10.225 while the approximate value is the area of the rectangle with base $10.225 - 10.215 = 0.01$ and height equal to the height of the Normal curve $f(x)$ at $x = 10.22$. Provided that measurements to two decimal places are sufficiently accurate that this is a good approximation for all relevant values of μ and σ, then the probability of observing x is approximately proportional to $f(x)$. Hence the likelihood is is proportional to $f(x)$. But if σ was very small (less than 0.02, say) then an interval of length 0.01 would cover a substantial fraction of the effective range of $f(x)$ and the likelihood defined to be proportional to $f(x)$ might not be appropriate.

Provided that the accuracy of measurement is sufficiently high and does not depend on the values of the parameters (within the range that might reasonably have produced the data), the probability of the data, and hence the likelihood function, can be taken to be proportional to the joint probability density of the data. For continuous models, therefore, the standard rule is to replace the probability of the data by its joint probability density. But this is an approximation rather than a definition. If the measurements are sufficiently coarse this approximation may not be good enough.

A great advantage of the likelihood function is that it provides a basis for combining different types of data. For example, one may have counts of numbers of tracks for several crystal grains, measurements of lengths and angles of other tracks, and even some "censored" observations where the actual measurements are not observed but partial information, such as an upper or lower bound, is recorded. If one can specify the joint probability of all of these data in terms of the parameters of interest, then the likelihood function will automatically combine the information appropriately. This aspect is an important feature of the combined likelihood in Section 9.2 and also for fitting models to annealing data (Section 9.6) where there is informative censoring.

A.10.2 Maximum likelihood estimates

The likelihood function is used both for estimation and hypothesis testing. The *maximum likelihood estimates* of the parameters are those values for which the probability of the data, or equivalently L, is maximised. They are the values of the parameters for which the observed data are most probable.

There is a well-established theory of the properties of maximum likelihood estimates. Under very general conditions they are known to be *efficient*, which means, roughly, that they use the available data about as well as possible — in particular, other estimates are generally not as precise. Also there is a routine method for calculating their standard errors and approximate confidence intervals when L is a well-behaved function. "Well-behaved" means, roughly, that L has a well-defined maximum at a point inside the parameter space and is approximately parabolic in shape for parameter values near this point. This method involves calculating the matrix of second derivatives of $-L$ with respect to each parameter, and then inverting this matrix. An approximate standard error of an estimated parameter is given by the square root of the relevant diagonal element of the resulting matrix. An approximate 95% confidence interval can then be calculated as "estimate ± two standard errors".

For the example of a Poisson count over an area A, where L is given by (A.16) this method in fact leads to the approximate standard error given by (A.5), i.e., the square root of the observed count. For $r = 20$ and $A = 1$ this gives an estimated standard error of $\sqrt{20} = 4.47$ and a two-sigma confidence interval for ρ of 11.06 to 28.94.

A.10.3 Likelihood ratio tests

The likelihood function provides a general method of hypothesis testing that is particularly useful for choosing between competing models.

Suppose there are two probability models M_1 and M_0 that may have generated the data, where M_0 is a special case of M_1. One calculates the maximum value of the log likelihood under each model, denoted by L_1 and L_0. Because M_0 is a special case, L_0 cannot be greater than L_1. The difference $L_1 - L_0$, or equivalently $W = 2(L_1 - L_0)$, is a measure of how much better the data are supported by the more general model, M_1. The larger W is, the more evidence there is against M_0 and in favour of M_1. Under general conditions, if the data were in fact generated by M_0, the test statistic W has approximately a χ^2 distribution with degrees of freedom equal to the *extra* number of parameters fitted in M_1 compared with M_0. Such tests are called *generalised likelihood ratio* tests because W is a function of the ratio of the maximised likelihoods under the two models.

For example, suppose the data are two counts r_1 and r_2 made over areas A_1 and A_2, respectively, and the model M_1 is that these are independent counts from Poisson distributions with means $A_1\rho_1$ and $A_2\rho_2$, where ρ_1 and ρ_2 are unknown. Suppose we wish to test the hypothesis that $\rho_1 = \rho_2$. Then the special case model M_0 says that the two Poisson means are $A_1\rho$ and $A_2\rho$, where ρ is unknown. The reader may verify, by writing down the maximum likelihood estimates under each model and substituting them into the log likelihood, that

$$
\begin{aligned}
W &= 2(L_1 - L_0) \\
&= 2\left\{ r_1 \log\left(\frac{r_1}{A_1}\right) + r_2 \log\left(\frac{r_2}{A_2}\right) - (r_1 + r_2)\log\left(\frac{r_1 + r_2}{A_1 + A_2}\right) \right\}.
\end{aligned}
$$

A large value of W is evidence against hypothesis that $\rho_1 = \rho_2$, where "large" is judged in relation to the χ^2 distribution with 1 degree of freedom, because one more parameter is fitted in M_1 compared with M_0.

For example, for data $r_1 = 20$, $A_1 = 1$ and $r_2 = 28$, $A_2 = 2$ we find that $W = 1.448$. The p-value is the probability that a random value from a $\chi^2(1)$ distribution exceeds 1.448, which is 0.23, indicating that there is no evidence that ρ_1 and ρ_2 differ.

In this case there are other ways of testing the hypothesis that $\rho_1 = \rho_2$. But the W method applies in just the same way to much more complex models and hypotheses.

A.10.4 Likelihood-based confidence intervals

Approximate confidence intervals may also be obtained directly from the log likelihood function. This is better than the method described in Section A.10.2 when L is not parabolic in shape near its maximum. The method is illustrated in Figure A.2 for the case of a single parameter. The figure shows a vertical line drawn down from the maximum of L to indicate the maximum likelihood estimate of ρ, which is $\hat{\rho} = 20$. Also a horizontal line is drawn at a height of 1.92 units below the maximum log likelihood which meets the log likelihood at two points. The values of ρ having a log likelihood within 1.92 units of the maximum (i.e., between the two other vertical lines) define an approximate 95% confidence interval. In this case the confidence interval for ρ is $(12.47, 30.09)$.

It is interesting to compare this interval with the so-called "exact" interval from Pearson and Hartley (1966, Table 40), which is $(12.22, 30.89)$. The latter is very slightly more conservative, as is to be expected from theory. Both of these differ a bit from the interval $(11.06, 28.94)$ found in Section A.10.2, which is slightly biased because L is not quite symmetrical for ρ near 20.

The rationale for this method derives from inverting a series of significance tests (see Section A.4). For each possible value of ρ the null hypothesis that that is the true value is tested. The confidence interval comprises those values of ρ for which the p-value is less than 0.05. The test statistic is $W = 2(L_{\max} - L)$ where L is the log likelihood corresponding to the hypothesised ρ and $L_{\max}$ is the maximum log likelihood. As implied in Section A.10.3, if the hypothesised ρ is the true value then W is from a distribution that is approximately χ^2 with 1 degree of freedom. (Here the model M_1 is that the observed count is from a Poisson distribution with unknown mean ρ while M_0 is the special case where ρ is fixed at its hypothesised value.) Then the p-value will be less than 0.05 if W is less than 3.84, the upper 5 percentage point of the $\chi^2(1)$ distribution, or equivalently when $L_{\max} - L$ is less than 1.92. This leads to the method illustrated in Figure A.2.

If there is more than one parameter, and a confidence interval for just one parameter is wanted, then this method may be applied using the *profile* log likelihood function for that parameter. The profile log likelihood function for a single parameter β, say, is the log likelihood function of β obtained by successively fixing β at different values over its range and at each value, maximising L over all other parameters. That is, it is the maximum L for a given β as a function of β.

Figure 6.5 shows profile log likelihood functions for each of the four parameters in the minimum age model, along with approximate 95% confidence intervals. For example, to calculate the profile for the minimum age, the parameter γ is fixed and the maximum L over possible values of μ, σ and π is calculated. This is repeated for a range of values of γ. The third panel in Figure 6.5 shows these maximum values of L plotted against γ — or rather against the age in Ma corresponding to γ, for ease of interpretation. Similarly for the other parameters. Note that the profile for σ is rather asymmetric, suggesting that the method described in Section A.10.2 might not be adequate.

This method can be extended to obtain joint confidence regions for two or more parameters.

References

Agresti, A. (1996) *An Introduction to Categorical Data Analysis.* Wiley, New York.

Aitken, M.J. (1990) *Science-Based Dating in Archaeology.* Longman, London.

Anderson, D.A. (1988) Some models for overdispersed binomial data. *Australian Journal of Statistics,* **30,** 125–148.

Anscombe, F.J. (1948) The transformation of Poisson, binomial and negative-binomial data. *Biometrika,* **35,** 246–254.

Anscombe, F.J. (1950) Sampling theory of the negative binomial and logarithmic series distributions. *Biometrika,* **37,** 358–382.

Armitage, P., Berry, G. and Matthews, J.N.S. (2002) *Statistical Methods in Medical Research.* (4th ed.) Blackwell, Oxford.

Arne, D.C. (1992a) The application of fission track thermochronology to the study of ore deposits. In: Zentilli, M. and Reynolds, P.H. (Eds.) *Short Course Handbook on Low Temperature Thermochronology.* Toronto: Mineralogical Society of Canada, **20,** 21–42.

Arne, D.C. (1992b) Evidence from apatite fission track analysis for regional Cretaceous cooling in the Ouchita Mountain fold belt and Arkoma Basin of Arkansas. *American Association of Petroleum Geologists Bulletin,* **76,** 392–402.

Barbarand, J., Hurford, A.J. and Carter, A. (2003a) Variation in apatite fission-track length measurement: implications for thermal history modelling. *Chemical Geology,* **198,** 77–106.

Barbarand, J., Carter, A., Wood, I. and Hurford, A.J. (2003b) Compositional and structural control of fission track annealing in apatite. *Chemical Geology,* **198,** 107–137.

Bigazzi, G., Bonadonna, F.P., Laurenzi, M.A. and Toranini, S. (1993) A test sample for FT dating of glass shards. *Nuclear Tracks,* **21,** 489–497.

Bock, R.D. and Lieberman, M. (1970) Fitting a response model for n dichotomously scored items. *Psychometrika,* **35,** 179–197.

Box, G.E.P., Hunter, W.J. and Hunter, J.S. (1978) *Statistics for Experimenters: An Introduction to Design, Data Analysis and Model Building.* Wiley, New York.

Brandon, M.T. (1992) Decomposition of fission-track grain-age distributions. *American Journal of Science,* **292,** 535–564.

Brandon, M.T. (1996) Probability density plot for fission-track grain-age samples. *Radiation Measurements,* **26,** 663–676.

Burchart, J., Dakowski, M. and Galazka, J. (1975) A technique to determine extremely high fission track densities. *Bulletin de l'Academie Polonaise des Sciences, Série de Science de la Terre,* **23,** 1–7.

Carlson, W.D., Donelick, R.A. and Ketcham, R.A. (1999) Variability of apatite fission-track annealing kinetics: I. Experimental results. *American Mineralogist,* **84,** 1213–1223.

Coleman, R. (1972) Sampling procedures for the lengths of random straight lines. *Biometrika*, **59**, 415–426.

Coleman, R. (1974) The distance from a given point to the nearest end of one member of a random process of linear segments. In: Harding, E.F. and Kendall, D.G. (Eds.) *Stochastic Geometry: a Tribute to the Memory of Rollo Davidson*. Wiley, London, 192–201.

Cowan, R.J. (1979) Homogeneous line segment processes. *Mathematical Proceedings of the Cambridge Philosophical Society*, **86**, 481–489.

Cox, D.R. (1969) Some sampling problems in technology. In: Johnson, N. and Smith, H. (Eds.) *New Developments in Survey Sampling*. Wiley, New York, 506–527.

Cox, D.R. (1970) *The Analysis of Binary Data*. Methuen, London.

Cox, D.R. and Isham, V.S. (1980) *Point Processes*. Chapman & Hall, London.

Crowley, K.R., Cameron, M. and Schaefer, R.L. (1991) Experimental studies of annealing of etched fission tracks in fluorapatite. *Geochimica et Cosmochimica Acta*, **55**, 1449–1465.

CYTEL Software Corporation (1991) StatXact: Statistical Software for Exact Non-parametric Inference, User Manual Version 2. CYTEL Software Corporation, Cambridge, Massachusetts.

Dakowski, M. (1978) Length distributions of fission tracks in thick crystals. *Nuclear Track Detection*, **2**, 181–190.

Dickin, A.P. (2004) *Radiogenic Isotope Geology*. (2nd ed.) Cambridge University Press, New York.

Donelick, R.A., Roden, M.K., Mooers, J.B., Carpenter, B.S. and Miller, D.S. (1990) Etchable length reduction of induced fission tracks in apatite at room temperature ($\sim 23°C$): crystallographic orientation effects and "initial" mean length. *Nuclear Tracks and Radiation Measurements*, **15**, 261–265.

Donelick, R.A. (1991) Crystallographic orientation dependence of mean etchable fission track length in apatite: an empirical model and experimental observations. *American Mineralogist*, **76**, 83–91.

Donelick, R.A., Ketcham, R.A. and Carlson, W.D. (1999) Variability of apatite fission-track annealing kinetics: II. Crystallographic orientation effects. *American Mineralogist*, **84**, 1224–1234. Erratum (2000): *American Mineralogist*, **85**, 1565.

Duddy, I.R., Green, P.F. and Laslett, G.M. (1988) Thermal annealing of fission tracks in apatite, 3. Variable temperature annealing. *Chemical Geology (Isotope Geoscience Section)*, **73**, 25–38.

Dumitru, T.A., Hill, K.C., Coyle, D.A., Duddy, I.R., Foster, D.A., Gleadow, A.J.W., Kohn, B.P., Laslett, G.M. and O'Sullivan, A.J. (1991) Fission track thermochronology: application to continental rifting of south-eastern Australia. *Australian Petroleum Exploration Association Journal*, **31**, 131–142.

Dumitru, T.A. (2000) Fission track geochronology. In: Noller, J.S., Sowers, J.M. and Lettis, W.R. (Eds.) *Quaternary Geochronology: Applications to Quaternary Geology and Palaeo-seismology, American Geophysical Union Reference Shelf*, **4**, 131–156.

Durrani, S.A. and Bull, R.K. (1987) *Solid State Nuclear Track Detection: Principles, Methods and Applications*. Pergamon Press, Oxford.

Elliot, J.C. (1994) *The Structure and Chemistry of the Apatites and other Calcium Orthophosphates*. Elsevier, Amsterdam.

Everitt, B.S. and Hand, D.J. (1981) *Finite Mixture Distributions*. Chapman & Hall, London.

Fisher, R.A. (1936) Has Mendel's work been re-discovered? *Annals of Science*, **I**, 115–137.

Fleischer, R.L., Price, P.B. and Walker, R.M. (1975) *Nuclear Tracks in Solids*. University of California Press, Berkeley.

Foster, D.A., Miller, D.S. and Miller, C.F. (1991) Tertiary extension in the Old Woman Mountains area, California: evidence from apatite fission track analysis. *Geology*, **10**, 875–886.

Galbraith, R.F. (1981) On statistical models for fission track counts. *Journal of Mathematical Geology*, **13**, 471–478; Reply: 485–488.

Galbraith, R.F. (1984) On statistical estimation in fission track dating. *Journal of Mathematical Geology*, **16**, 653–669.

Galbraith, R.F. (1986) Statistical analysis of C.W. Naeser's Fish Canyon zircon data. *Nuclear Tracks and Radiation Measurements*, **11**, 295–300.

Galbraith, R.F. (1988) Graphical display of estimates having differing standard errors. *Technometrics*, **30**, 271–281.

Galbraith, R.F. and Laslett, G.M. (1988) Some calculations relevant to thermal annealing of fission tracks in apatite. *Proceedings of the Royal Society of London A* **419**, 305–321.

Galbraith, R.F. (1990) The radial plot: graphical assessment of spread in ages. *Nuclear Tracks and Radiation Measurements*, **17**, 207–214.

Galbraith, R.F. and Green, P.F. (1990) Estimating the component ages in a finite mixture. *Nuclear Tracks and Radiation Measurements*, **17**, 197–206.

Galbraith, R.F., Laslett, G.M., Green, P.F. and Duddy, I.R. (1990) Apatite fission track analysis: geological thermal history analysis based on a three-dimensional random process of linear radiation damage. *Philosophical Transactions of the Royal Society of London A*, **332**, 419–438.

Galbraith, R.F. and Laslett, G.M. (1993) Statistical models for mixed fission track ages. *Nuclear Tracks and Radiation Measurements*, **21**, 459–470.

Galbraith, R.F. (1994) Some applications of radial plots. *Journal of the American Statistical Association*, **89**, 1232–1242.

Galbraith, R.F. and Laslett, G.M. (1997) Statistical modelling of thermal annealing of fission tracks in zircon. *Chemical Geology (Isotope Geoscience Section)*, **140**, 123–135.

Galbraith, R.F. (1998) The trouble with probability density plots of fission track ages. *Radiation Measurements*, **29**, 125–131.

Gallagher, K. (1995) Evolving temperature histories from apatite fission-track data. *Earth and Planetary Science Letters*, **136**, 421–435.

Gallagher, K., Brown, R. and Johnson, C. (1998) Fission track analysis and its applications to geological problems. *Annual Reviews in Earth and Planetary Science*, **26**, 519–572.

Gleadow, A.J.W. (1978) Anisotropic and variable track etching characteristics in natural sphenes. *Nuclear Track Detection*, **2**, 105–117.

Gleadow, A.J.W. (1981) Fission track dating methods: what are the real alternatives? *Nuclear Tracks* **5**, 3–14.

Gleadow, A.J.W. and Duddy, I.R. (1981) A natural long term annealing experiment for apatite. *Nuclear Tracks* **5**, 169–174.

Gleadow, A.J.W., Duddy, I.R. and Lovering, J.F. (1983) Fission track analysis: a new tool for the evaluation of thermal histories and hydrocarbon potential. *Australian Petroleum Exploration Association Journal*, **23**, 93–102.

Gleadow, A.J.W., Duddy, I.R., Green, P.F. and Lovering, J.F. (1986) Confined fission track lengths in apatite: a diagnostic tool for thermal history analysis. *Contributions to Mineralogy and Petrology*, **94**, 405–415.

Goswami, J.N., Jha, R. and Lal, D. (1984) Quantitative treatment of annealing of charged particle tracks in common rock minerals. *Earth and Planetary Science Letters*, **71**, 120–128.

Goutis, C. (1993) Recovering extra-binomial variation. *Journal of Statistical Computation and Simulation*, **45**, 233–242.

Goutis, C. (1997) Non-parametric estimation of a mixing distribution via the kernel method. *Journal of the American Statistical Association*, **92**, 1445–1450.

Gradshteyn, I.S. and Ryzhik, I.M. (1980) *Tables of Integrals, Series and Products: Corrected and Enlarged Edition*. Academic Press, New York.

Green, P.F. and Durrani, S.A. (1977) Annealing studies of tracks in crystals. *Nuclear Tracks*, **1**, 33–39.

Green, P.F. (1981a) A new look at statistics in fission track dating. *Nuclear Tracks*, **5**, 77–86.

Green, P.F. (1981b) "Track-in-track" length measurements in annealed apatites. *Nuclear Tracks*, **5**, 121–128.

Green, P.F., Duddy, I.R., Gleadow, A.J.W., Tingate, P.R. and Laslett, G.M. (1985) Fission-track annealing in apatite: track length measurements and the form of the Arrhenius plot. *Nuclear Tracks*, **10**, 323–328.

Green, P.F. (1986) On the thermo-tectonic evolution of Northern England: evidence from fission-track analysis. *Geological Magazine*, **123**, 493–506.

Green, P.F., Duddy, I.R., Gleadow, A.J.W., Tingate, P.R. and Laslett, G.M. (1986) Thermal annealing of fission tracks in apatite, 1. A qualitative description. *Chemical Geology (Isotope Geoscience Section)*, **59**, 237–253.

Green, P.F. (1988) The relationship between track shortening and fission track age reduction in apatite. *Earth and Planetary Science Letters*, **89**, 335–352.

Green, P.F., Duddy, I.R. and Laslett, G.M. (1988) Can fission track annealing in apatite be described by first-order kinetics? *Earth and Planetary Science Letters*, **87**, 216–228.

Green, P.F. (1989) *UK Onshore Well Data*. Personal communication extracted from Geotrack Report 122, May 1989.

Green, P.F., Duddy, I.R., Gleadow, A.J.W. and Lovering, J.F. (1989a) Apatite fission track analysis as a paleotemperature indicator for hydrocarbon exploration. In: Naeser, N.D. and McCulloh, T.H. (Eds.) *Thermal History of Sedimentary Basins: Methods and Case Histories*. Springer-Verlag, New York, 181–195.

Green, P.F., Duddy, I.R., Laslett, G.M., Hegarty, K.A., Gleadow, A.J.W. and Lovering, J.F. (1989b) Thermal annealing of fission tracks in apatite, 4. Quantitative modelling techniques and extension to geological timescales. *Chemical Geology (Isotope Geoscience Section)*, **79**, 155–182.

Greig-Smith, P. (1952) The use of random and contiguous quadrats in the study of the structure of plant communities. *Annals of Botany*, **16**, 293–316.

Hayashi, M. (1985) A statistical method for eliminating the data pertaining to detrital grains in fission track dating (abstract). *Nuclear Tracks and Radiation Measurements*, **10**, 414.

Hurford, A.J. and Green, P.F. (1983) The zeta age calibration of fission-track dating. *Isotope Geoscience*, **1**, 285–317.

Hurford, A.J., Fitch, F.J. and Clarke, A. (1984) Resolution of the age structure of the detrital zircon populations of two Lower Cretaceous sandstones from the Weald of England by fission track dating. *Geological Magazine*, **121**, 269–277.

Hurford, A.J. (1991) Uplift and cooling pathways derived from fission track analysis and mica dating: a review. *Geologische Rundschau*, **80**, 349–368.

Hurford, A.J. and Carter, A. (1991) The role of fission track dating in the discrimination of provenance. In: Morton, A.C., Todd, S.P. and Haughton, P.D.W. (Eds.) *Developments in Sedimentary Provenance Studies*. Geological Society of London Special Publications, **57**, 67–78.

Ishikawa, N. and Tagami, T. (1991) Paleomagnetism and fission track geochronology on the Goto and Tsushima Islands in the Tsushima Strait area: implications for the opening mode of the Japan Sea. *Journal of Geomagnetism and Geoelectricity*, **43**, 229–253.

Issler, D.R., Beaumont, C., Willett, S.D., Donelick, R.A., Mooers, J. and Grist, A. (1990) Preliminary evidence from apatite fission track data concerning the thermal history of the Peace River Arch Region, Western Canada Sedimentary Basin. *Bulletin of Canadian Petroleum Geology*, **38A**, 250–269.

Johnson, N.L. and Leone, F.C. (1964) *Statistics and Experimental Design: in Engineering and the Physical Sciences, Volume II*. Wiley, New York.

Kallmes, O., Corte, H. and Bernier, G. (1961) The structure of paper II. The statistical geometry of a multiplanar fibre network. *Technical Association of the Pulp and Paper Industry*, **44**, 519–528.

Kamp, P.J.J. and Green, P.F. (1991) Thermal and tectonic history of selected Taranaki Basin (New Zealand) wells assessed by apatite fission track analysis. *American Association of Petroleum Geologists Bulletin*, **74**, 1401–1419.

Ketcham, R.A., Donelick, R.A. and Carlson, W.D. (1999) Variability of apatite fission-track annealing kinetics: III. Extrapolation to geological time scales. *American Mineralogist*, **84**, 1235–1255.

Kowallis, B.J., Heaton, J.S. and Bringhurst, K. (1986) Fission-track dating of volcanically derived sedimentary rocks. *Geology*, **14**, 19–22.

Laird, N.M. (1978) Non-parametric maximum likelihood estimation of a mixing distribution. *Journal of the American Statistical Association*, **73**, 805–811.

Laslett, G.M., Kendall, W.S., Gleadow, A.J.W. and Duddy, I.R. (1982) Bias in measurement of fission track length distributions. *Nuclear Tracks*, **6**, 79–85.

Laslett, G.M., Gleadow, A.J.W. and Duddy, I.R. (1984) The relationship between fission track length and track density in apatite. *Nuclear Tracks*, **9**, 29–38.

Laslett, G.M., Green, P.F., Duddy, I.R. and Gleadow, A.J.W. (1987) Thermal annealing of fission tracks in apatite, 2. A quantitative analysis. *Chemical Geology (Isotope Geoscience Section)*, **65**, 1–13.

Laslett, G.M. (1993) Modelling the conditional length distribution of fully etched horizontal confined fission tracks in apatite: a proposal. *CSIRO Division of Mathematics and Statistics Report*, DMS-E93/79, 21pp.

Laslett, G.M., Galbraith, R.F. and Green, P.F. (1994) The analysis of projected fission track lengths. *Radiation Measurements*, **23**, 103–123.

Laslett, G.M. and Galbraith, R.F. (1996a) Statistical properties of semi-tracks in fission track analysis. *Radiation Measurements* **26**, 565–576.

Laslett, G.M. and Galbraith, R.F. (1996b) Statistical modelling of thermal annealing of fission tracks in apatite. *Geochimica et Cosmochimica Acta*, **60**, 5117–5131.

Laslett, G.M. and Galbraith, R.F. (2000) Statistical modelling of thermal annealing of fission tracks in Durango apatite. *Research Report No. 217*, Department of Statistical Science, University College London.

Lehman, R.L. and Brisbane, R.W. (1968) Random-drift sampling; a study by computer simulation. *Nuclear Instruments and Methods*, **64**, 269–277.

McGee, V.E. and Johnson, N.M. (1979) Statistical treatment of experimental errors in the fission track dating method. *Journal of Mathematical Geology*, **11**, 255–268.

McGee, V.E., Johnson, N.M. and Naeser, C.W. (1985) Simulating the fission track dating procedure. *Nuclear Tracks*, **10**, 365–379.

McLachlan, G.J. and Peel, D. (2000) *Finite Mixture Models*. Wiley, New York.

Miller, D.S., Crowley, K.D., Dokka, R.K., Galbraith, R.F., Kowallis, B.J. and Naeser, C.W. (1993) Results of interlaboratory comparison of fission track ages for the 1992 Fission Track Workshop. *Nuclear Tracks and Radiation Measurements*, **21**, 565–573.

Moore, M.E., Gleadow, A.J.W. and Lovering, J.F. (1986) Thermal evolution of rifted continental margins: new evidence from fission tracks in basement apatites from south eastern Australia. *Earth and Planetary Science Letters*, **78**, 255–270.

Naeser, C.W. (1979a) Fission track dating and geological annealing of fission tracks. In: Jäger, E. and Hunziker, J.C. (Eds.) *Lectures in Isotope Geology*. Springer-Verlag, New York, 154–169.

Naeser C.W. (1979b) Thermal history of sedimentary basins by fission track dating of sub-surface rocks. In: Scholle P.A., Schluger, P.R. (Eds.) *Aspects of Diagenesis*. Society of Economic Paleontology and Mineralogy Special Publications, **26**, 109–112.

Naeser, N.D. and Naeser C.W. (1984) Fission track dating. In: Mahaney, W.C. (Ed.) *Quaternary Dating Methods*. Elsevier, New York, 87–100.

Naeser, N.D., Zeitler, P.K., Naeser, C.W. and Cerveny, P.F. (1987) Provenance studies by fission track dating of zircon — etching and counting procedures. *Nuclear Tracks and Radiation Measurements*, **13**, 121–126.

Ogston, A.G. (1958) The spaces in a uniform random suspension of fibres. *Transactions of the Faraday Society*, **54**, 1754–1757.

Ogston, A.G., Preston, B.N. and Wells, J.D. (1973) On the transport of compact particles through solutions of chain polymers. *Proceedings of the Royal Society of London A*, **333**, 297–316.

Omar, G.I., Steckler, M.S., Buck, W.R. and Kohn, B.P. (1989) Fission track analysis of basement apatites at the Western Margin of the Gulf of Suez Rift, Egypt: evidence for synchronicity of uplift and subsidence. *Earth and Planetary Science Letters*, **94**, 255–270.

Parker, P. and Cowan, R. (1976) Some properties of line segment processes. *Journal of Applied Probability*, **13**, 96–107.

Paul, T.A. and Fitzgerald, P.G. (1992) Transmission electron microscope investigation on fission tracks in fluorapatite. *American Mineralogist*, **77**, 336–344.

Pearson, E.S. and Hartley, H.O. (1966) *Biometrika Tables for Statisticians, Volume I*. (3rd ed.) Cambridge University Press, Cambridge.

Price, P.B. and Walker, R.M. (1963) Fossil tracks of charged particles in mica and the age of minerals. *Journal of Geophysical Research*, **68**, 4847–4862.

Ravenhurst, C.E., Roden-Tice, M.K. and Miller, D.S. (2003) Thermal annealing of fission tracks in fluorapatite, chlorapatite, mangamoan apatite and Durango apatite: experimental results. *Canadian Journal of Earth Sciences*, **40**, 995–1007.

Rice, J.A. (1995) *Mathematical Statistics and Data Analysis*. (2nd ed.) Duxbury Press, Belmont.

Ripley, B.D. (1981) *Spatial Statistics*. Wiley, New York.

Ross, S.M. (1997) *Introduction to Probability Models*. (6th ed.) Academic Press, San Diego.

Sandhu, A.S., Westgate, J.A. and Alloway, B.V. (1993) Optimising the isothermal plateau fission-track dating method for volcanic glass shards. *Nuclear Tracks and Radiation Measurements*, **21**, 479–488.

Sandhu, A.S. and Westgate, J.A. (1995) The correlation between reduction in fission track diameter and areal track density in volcanic glass shards and its application in dating tephra beds. *Earth and Planetary Science Letters*, **131**, 289–299.

Schwartz, G. (1978) Estimating the dimension of a model. *Annals of Statistics*, **6**, 461–464.

Seward, D. and Rhoades, D.A. (1986) A clustering technique of fission track dating of fully to partially annealed minerals and other non-unique populations. *Nuclear Tracks and Radiation Measurements*, **11**, 259–268.

Silk, E.C.H. and Barnes, R.S. (1959) Examination of fission fragment tracks with an electron microscope. *Philosophical Magazine*, **4**, 970–971.

Snedecor, G.W. and Cochran, W.G. (1980) *Statistical methods*. (7th ed.) Iowa State University Press, Ames.

Sobel, E.R. and Dumitru, T.A. (1997) Thrusting and exhumation around the margins of the western Tarim basin during the India-Asia collision. *Journal of Geophysical Research*, **102**, 5043–5064.

Statistical Sciences (1992) *S-plus User's Manual, Version 3.1*. Statistical Sciences Inc., Seattle, Washington.

Storzer, D. and Wagner, G.A. (1969) Correction of thermally lowered fission track ages of tektites. *Earth and Planetary Science Letters*, **5**, 463–468.

Storzer, D. and Poupeau, G. (1973) Ages-plateaux de minéreaux et verres par la méthods des traces de fission. *Comptes Rendus de l'Academie des Sciences, Paris*, **276**, 137–139.

Tagami, T., Galbraith, R.F., Yamada, R. and Laslett, G.M. (1998) Revised annealing kinetics of fission tracks in zircon and geological implications. In: Van den haute, P. and De Corte, F. (Eds.) *Advances in Fission-Track Geochronology*. Kluwer Academic Publishers, Dordrecht, 99–112.

Tippett, J.M. and Kamp, P.J.J. (1993) Fission track analysis of the Late Cenozoic vertical kinematics of continental Pacific crust, South Island, New Zealand. *Journal of Geophysical Research*, **98**, 16119–16148.

Titterington, D.M., Smith, A.F.M. and Makov, U.E. (1985) *Statistical Analysis of Finite Mixture Distributions*. Wiley, Chichester.

van der Touw, J.W., Galbraith, R.F. and Laslett, G.M. (1997) A logistic truncated normal mixture model for overdispersed binomial data. *Journal of Statistical Computation and Simulation*, **59**, 349–373.

Wagner, G.A. (1979a) Archaeometric dating. In: Jäger, E. and Hunziker, J.C. (Eds.) *Lectures in Isotope Geology*. Springer-Verlag, New York, 178–188.

Wagner, G.A. (1979b) Correction and interpretation of fission track ages In: Jäger, E. and Hunziker, J.C. (Eds.) *Lectures in Isotope Geology*. Springer-Verlag, New York, 170–177.

Wagner, G. and Van den haute, P. (1992) *Fission Track Dating*. Kluwer, Dordrecht.

Watt, S. and Durrani, S.A. (1985) Thermal stability of fission tracks in apatite and sphene: using confined-track-length measurements. *Nuclear Tracks* **10,** 349–357.

Westgate, J.A. (1989) Isothermal plateau fission-track ages of hydrated glass shards from silicic tephra beds. *Earth and Planetary Science Letters,* **95,** 858–861.

Williams, D.A. (1982) Extra-binomial variation in logistic linear models. *Applied Statistics,* **31,** 144–148.

Yamada, R., Tagami, T., Nishimura, S. and Ito, H. (1995) Annealing kinetics of fission tracks in zircon: an experimental study. *Chemical Geology (Isotope Geoscience Section),* **122,** 249–258.

Index

Mathematical notation

$E[l|\phi]$, mean length of tracks at angle ϕ to c-axis, 16

$f(l)$, probability density function of track length, 16

$f_\phi(l)$, probability density function of lengths of tracks at angle ϕ, 16

I, isotopic ratio, 8, 30

l, track length, 15

r, projected semi-track length, 115

r, length reduction factor, 172

T, absolute temperature, 171

t, fission track age, 28

t, semi-track length, 115

t, time elapsed, 171

t, time point, 176